Crew Leadership

Introductory Skills for the Crew Leader

This information is general in nature and intended for training purposes only. Actual performance of activities described in this manual requires compliance with all applicable operating, service, maintenance, and safety procedures under the direction of qualified personnel. References in this manual to patented or proprietary devices do not constitute a recommendation of their use.

ISBN 0-13-103593-

TABLE OF CONTENTS

CHAPTER TWO–LEADERSHIP SKILLS

LIST OF FIGURES

ACKNOWLEDGMENTS

The NCCER wishes to acknowlege the dedication and expertise of Roger Liska, the original author and mentor for this series on leadership development.

Roger W. Liska, Ed.D., FCIOB, CPG, FAIC, PE

Clemson University

Chair & Professor

Department of Construction Science & Management

We would also like to thank the following reviewers for contributing their time and expertise to this endeavor:

J.R. Blair

Tri-City Electrical Contractors

An Encompass Company

Mike Cornelius

Tri-City Electrical Contractors

An Encompass Company

Dan Faulkner

Wolverine Building Group

David Goodloe

Clemson University

Kevin Kett

The Haskell Company

Danny Parmenter

The Haskell Company

FOREWORD

The crew leader plays a key role in the operation of a construction company. As the first line of management on the job, the crew leader is responsible for supervising workers in an effective and efficient manner. Once a construction worker is promoted to the position of crew leader, he or she becomes involved with managing people.

As a supervisor, the crew leader will use new skills. These skills include problem solving, planning, scheduling, estimating, controlling the use of job resources, such as material and equipment, and knowing how to perform the work in a safe manner. Practicing effective human relations skills, such as motivating workers and communicating with them, are also an essential part of the crew leader's job.

The purpose of this program is to provide the person who is considering becoming a crew leader with knowledge of the basic leadership skills required for the job.

This program consists of an Instructor's Guide and a Participant's Manual. The Instructor's Guide contains a breakdown of the information provided in the Participant's Manual as well as the actual text that the participant will use. The Participant's Manual contains the material that the participant will study, along with self-check exercises and activities, to evaluate whether the participant has mastered the knowledge necessary to become an effective crew leader. This course is designed to be taught in eight (8) two-hour sessions.

For the participant to gain the most from this program, it is recommended that the material be presented in a formal classroom setting, using a trained and experienced instructor. If the student is so motivated, he or she can study the material on a self-learning basis by using the material in both the Manual and Guide. Recognition through the National Registry is available for the participants provided the program is delivered through the Accredited Sponsor network by a Master Trainer or ICTP graduate. More details on this program can be received by calling the National Center for Construction Education and Research at 352-334-0911.

Users of this program should note that some examples provided to reinforce the material may not apply to the participant's exact work, although the process does. Furthermore, every company has its own mode of operation. Therefore, some topics may not apply to the participant's company. Topics have been included because they are important considerations for prospective crew leaders throughout the industry.

A NOTE TO NCCER INSTRUCTORS AND TRAINEES

If you are training under an Accredited NCCER Sponsor, note that you may be eligible for dual credentials for successful completion of Introductory Skills for the Crew Leader. When submitting the Form 200, indicate completion of the two module numbers that apply to Introductory Skills for the Crew Leader - MT101 (from NCCER's Management series) and 27409-03 (from NCCER's Carpentry Level Four) and transcripts will be issued to you accordingly.

NATIONAL
CENTER FOR
CONSTRUCTION
EDUCATION AND
RESEARCH

CHAPTER ONE

Orientation to the Job

OBJECTIVES

Upon completion of this chapter, you should be able to:

1. Discuss the history, trends, and economic conditions of the construction industry.

2. Describe how workers' values have changed over the years.

3. Explain the importance of training for construction industry personnel.

4. List the new technologies available, and discuss how they are helpful to the construction industry.

5. Identify the gender and minority issues associated with a changing workforce.

6. Describe what employers can do to prevent workplace discrimination.

7. Describe the four major categories of construction projects.

8. Differentiate between formal and informal organizations.

9. Describe the difference between authority and responsibility.

10. Explain the purpose of job descriptions and what they should include.

11. Distinguish between company policies and procedures.

SECTION 1

1.0.0 OVERVIEW OF THE CONSTRUCTION INDUSTRY

Today the construction industry is providing more than buildings where people live, work, shop, worship, and learn. It is also involved in the building of other structures, such as roads, bridges, airports, and tunnels that enable goods and people to move freely about the country. In addition, the construction industry provides dams for water and power, as well as sewage systems, all with one goal: to make our lives more comfortable.

The construction industry is one of the largest industries in America. It is larger than the steel and auto industries combined. In recent years the total value of new construction exceeded $500 billion.

1.1.0 Historical Importance of the Construction Industry

To fully understand the future needs of the construction industry, it is necessary to review the past. The history of construction comes before the earliest written records. It can be traced back to the need to build shelters against the elements and predators, as well as to raise great monuments to gods or honor the dead.

Builders of all ages have always depended on the availability of three elements: materials, labor, and technology. The pyramids of ancient Egypt are a good example of how early builders used these elements to create structures that satisfied their needs. The new English Channel Tunnel and the Mall of America outside Minneapolis, Minnesota represent the modern builders' use of materials, labor, and technology.

The construction industry will continue to change as engineers and architects search for new ways of using resources. Contractors will discover new ways of doing old jobs, and supervisors will continue to learn new technical and management methods.

1.2.0 Growth and Economics of the Construction Industry

Construction is the second largest industry in the United States, employing over 5 million workers. As the population of our country increases, so will the need for new construction. Aging infrastructure and new environmental laws will contribute to this increase. Finally, more and more American companies will be needed for construction overseas.

With this increased demand for construction comes a stronger need for trained craftworkers. The U.S. Department of Labor (DOL) reported that employment in the construction industry grew 48 percent between January 1993 and September 2000, yet construction companies continue to have a shortage of skilled applicants.

The tight labor market and the boom in information technology, including computerization, have resulted in a decrease in available, qualified construction workers. The DOL reports that there is a shortage of approximately 240,000 workers per year in the construction industry. Because of the labor shortage, construction projects across the country are being delayed, particularly residential construction jobs.

To compensate for the lack of skilled construction labor, companies are being forced to change

their approach to recruitment. A spokesperson for the DOL said, "Employers need to rethink their mindsets about who are qualified workers." Research suggests that the construction industry is branching out in an attempt to recruit and train younger workers just out of high school and is reaching out to women and minorities.

1.3.0 Changing Values of Workers

The construction industry can no longer expect that workers will simply show up and stay throughout a project. Instead, extra efforts must be made to attract and retain a skilled workforce. To do so, construction companies must be aware of what workers value.

Because the unemployment rate is so low, it is definitely a laborer's job market. Therefore, construction companies have to try even harder to attract potential workers to learn a trade.

Workers value a number of different things. Examples include recognition and praise, a sense of accomplishment, and opportunity for advancement. What an individual values and the extent to which it is valued varies from person to person.

Research conducted by the Society for Human Resources Management (SHRM) suggests that workers are not as committed to their organizations as they were in the past. The opportunities to change jobs are plentiful, so employees rarely work at one company throughout their careers as they did in past generations. Today, workers select companies where they can learn and develop new skills. Once these skills are developed and there is nothing more to learn, the workers move on to another job.

Training and education, the use of new technologies, and opportunities for advancement and growth within the construction profession are valuable sales tools being used by the construction industry to attract job candidates. These three areas can help to retain skilled workers for longer, as well.

REVIEW QUESTIONS ◆ CHAPTER ONE, SECTIONS 1.0.0–1.3.0

1. The construction industry has changed over the years in which of the following ways?

   a. The scope of structures being built has narrowed.

   b. The industry has become one of the largest, though not as large as the steel or auto industries.

   c. Early builders were not able to make use of technology.

   d. New methods of management now exist.

2. One of the reasons for changes in the construction industry is that _____.

   a. many resources are used differently than in the past

   b. employment in the construction industry has not grown in recent years

   c. there is less demand for American companies to construct in other countries

   d. the demand for construction is decreasing

3. There is a shortage of skilled workers in the construction industry today because _____.

   a. there is poor long-term job security in the construction industry

   b. recruiters have recently become more selective when searching for potential workers

   c. the labor market is tight

   d. technology has replaced the need for many construction jobs, which discourages potential workers

4. One way in which workers' values have changed over the years is that _____.

   a. workers expect more time off now than they did in the past

   b. employees value learning more now than they did in the past

   c. workers want a secure job environment in which to settle down and feel comfortable

   d. employees typically value a high salary above everything

5. Workers tend to be less loyal to their companies now then they were previously because _____.

   a. they typically feel their job is too difficult

   b. they desire challenges

   c. their job involves too much change and variety

   d. they feel their job is too physically dangerous

6. Construction companies are actively recruiting _____.

   a. retired people who would like to resume working

   b. students who have just finished college

   c. high school students who would like to work as apprentices

   d. students who have just completed high school

7. The following are some ways that employers are attempting to attract and retain skilled workers *except* _____.

   a. offering greater salaries

   b. providing growth opportunities

   c. leveraging technology

   d. making training available

SECTION 2

2.0.0 THE CONSTRUCTION INDUSTRY TODAY

There have been many changes in the construction industry over the years. The continued need for training and education, advances in technology, and the addition of women and minorities to the workforce have affected the way that construction companies operate.

2.1.0 Training

The need for continuing craft training is necessary if the industry is to meet the forecasted worker demands. The Construction Labor Research Council estimates that the construction industry will need more than 240,000 new workers annually for the year 2001 and beyond. Replacement needs will exceed growth demands by 250 percent.

Because older workers in construction are retiring earlier and younger workers are leaving construction for jobs in other industries, the construction industry will be forced to increase its share of the new job entrants to meet future demands. The Department of Labor (DOL) concludes that the best way for the construction industry to reduce shortages of skilled workers is to create more education and training opportunities. The DOL suggests that companies and community groups form partnerships and create apprenticeship programs. Such programs could provide women, young people, and minorities an opportunity to develop the skills needed for construction work by giving them hands-on experience.

Present construction craft apprenticeship and task-training programs are graduating an average of 50,000 people per year. If the training of construction workers is not significantly increased, a severe shortage of skilled help could result. Therefore, the two major concerns of the construction industry today are:

- Developing adequate training programs to provide skilled craftworkers to meet future needs

- Providing supervisory training to ensure there are qualified leaders in the industry to supervise the craftworkers

2.1.1 Craft Training

Most of the craft training done in the construction industry is rather informal, taking place on the job site, outside of a traditional training classroom. According to the American Society for Training and Development, craft training is generally handled through on-the-job instruction by a co-worker or conducted by a supervisor.

The Society of Human Resources Management (SHRM) offers the following tips to supervisors in charge of training their employees:

1. *Establish career goals for each person in your area.* This information will give you an idea of how much time you need to devote to developing each person. Employees with the highest ambitions and those with the lowest skill levels will require the most training.

2. *Determine what kind of training to give.* Training can be on the job under the supervision of a co-worker, it can be one-on-one with the supervisor, it can involve cross training to teach a new trade or skill, or it can involve delegating new or additional responsibilities.

3. *Determine the trainee's preferred method of learning.* Some people learn best by watching, others from verbal instructions, and others from doing. When training more than one person at a time, use all three methods.

The SHRM advises that supervisors do the following when training their employees:

1. *Explain the task, why it needs to be done, and how it should be done.* Confirm that the trainees understand these three areas by asking questions. Allow them to ask questions as well.

2. *Demonstrate the task.* Break the task down into manageable chunks. Do one part at a time.

3. *Ask trainees to do the task while you observe them.* Try not to interrupt them while they are doing the task unless they are doing something that is unsafe and potentially harmful.

4. *Give the trainees feedback.* Be specific about what they did and mention any areas where they need to improve.

2.1.2 Supervisory Training

Given the increased demand for skilled craft-workers and qualified supervisory personnel, it seems logical that training would be on the rise. While some contractors have developed their own in-house training program and others participate in training offered by associations and other organizations, many others do not offer training at all.

The reasons for not developing training programs include the following:

- Lack of money to train
- Lack of time to train
- Lack of knowledge about training programs
- High rate of employee turnover
- Workforce too small
- Past involvement ineffective
- Hiring only trained workers
- Lack of interest from workers

For craftworkers to move up into supervisory and managerial positions, it will be necessary for them to continue their education and training. Those who are willing to acquire and develop new skills will have the best chance of finding stable employment. It is, therefore, critical that construction companies remove the barriers against training and devote the needed resources to training.

Your company has recognized the need for training. Participating in a leadership training program such as this will begin to fill the gap between craft and supervisory training.

2.2.0 New Technology

Like many industries, the construction industry has made the move to technology as a means of remaining competitive. The inventions of the Internet, e-mail (electronic mail), the World Wide Web, e-commerce (electronic commerce), and other forms of communication have changed the way organizations do business in the 21st century.

In 1999, the architecture, engineering, and construction (AEC) industry conducted a survey on the current and planned use of information technology and its impact. The findings suggested that a majority of business processes, such as bookkeeping, invoicing, and specifica-tion writing are computerized, and the remaining business processes are likely to be computerized in the near future.

The firms surveyed reported that they planned to continue to increase their information technology investments because of the benefits that resulted. Benefits cited by survey respondents included increased productivity and speed, improved quality of documents, greater access to common data, and better financial controls and communication. As technnology becomes a greater part of supervision, crew leaders will need to be able to use it properly.

2.2.1 The Internet

The *Internet* refers to individual computers and networks or groups of computers that are linked together to share information. It was developed in the mid-1960s by the Department of Defense's Advanced Research Project Agency (ARPA). Known as the ARPANET, its purpose was to provide a means of communicating in military emergencies. In the beginning, only government "think tanks" and a few universities were linked.

Now, billions of computers around the world are linked via the Internet. The Internet provides its users the ability to access information and transmit information to others electronically. Internet users can research countless topics and display their own web pages on the World Wide Web, communicate with others via e-mail, and purchase items electronically.

2.2.2 E-Mail

E-mail refers to mail that is typed into a computer, then sent electronically to a recipient via the Internet. In other words, e-mail is sent from one computer to another without paper and envelopes, postage stamps, or delivery time.

There are many benefits of using e-mail. First, the transfer of information is very quick, usually taking only moments for the e-mail to reach the recipient's inbox. Second, there are no postage charges or fees associated with sending e-mail. Next, e-mail allows the sender to send drawings, files, and documents as attachments which the recipient(s) can download to a personal computer for viewing, printing, and editing.

Finally, the sender can send the same information to multiple recipients without having to reproduce it many times; in other words, the sender simply addresses the e-mail to a list of recipients and they all receive the same information at the same time.

2.2.3 The World Wide Web

The *World Wide Web* is essentially an illustrated version of the Internet. It began in the late 1980s when physicist Dr. Tim Berners-Lee wrote a small computer program for his own personal use. This program allowed pages within his computer to be linked together using keywords.

In 1992, it was found that it was possible to link documents in different computers, as long as the computers were connected to the Internet; thus, the World Wide Web was created. Originally, web pages were only text-based. Today, they are much more sophisticated, containing graphics, sound, animation, and text.

The World Wide Web can be used to conduct research on various topics by simply conducting a search of keywords, such as *construction bids* or *project control*. The click of a few buttons can result in a list of web sites with information on the subject.

The more specific the keywords, the more narrow the search results. For instance, searching with the keyword *construction* results in hundreds of thousands of web pages, while searching for *construction economics* narrows the results considerably.

A *website* refers to the location of a web page or series of web pages on the World Wide Web that contain information related to the keywords being searched. There are countless web pages on the World Wide Web, with many new web sites appearing every day.

2.2.4 E-Commerce

Many organizations are creating web pages to advertise their companies and their services. For example, the construction industry now has thousands of web pages related to architectural firms, contractors, suppliers, economics, project management systems, and construction industry jobs.

A 2000 survey conducted by the CIT Group found that 60 percent of construction distributors have company web sites. In addition, 62 percent of those who did not have a company web site plan on developing one in 2001.

Some company web pages are merely informational. Others provide a platform for e-commerce, or buying and selling over the Internet. Through e-commerce it is possible for businesses to communicate electronically with other businesses. For example, a contractor can order 100 12-foot 2×4s from a web site that sells lumber.

The contractor can shop prices and compare the quality of items available through different vendors. A purchase order can be submitted to the vendor once the appropriate item is selected, and the contractor can indicate any special instructions, such as how and where the items requested should be shipped. All of this can be done on the Internet, without completing any paper forms or making telephone calls to vendors for price quotes.

Once the purchase request is submitted over the Internet, it is sent to the vendor, who then fills the order, charges the contractor accordingly, and ships the merchandise according to the contractor's specifications. Unless a product is unavailable or out-of-stock, the response time between receiving the order electronically and shipping the merchandise to the contractor is drastically reduced as a result of e-commerce.

E-commerce results in time and money benefits for both the contractor and the vendor. The contractor benefits from the ability to:

- Compare merchandise
- Order merchandise in a matter of a few clicks on the computer
- Receive the merchandise more quickly than using telephone, fax, or mail.

The vendor benefits because:

- Inventory can be computerized and monitored more accurately
- Orders can be filled more efficiently

2.2.5 Other Communication Sources

Cellular telephones, voicemail, pagers, and handheld communication devices have made it even easier to stay in touch. They are particularly useful communication sources for contractors or crew leaders who are on a job site, away from their offices, or constantly on the go.

Cellular telephones allow the users to receive incoming calls as well as make outgoing calls. Unless the owner is out of the cellular provider's service area, the cell phone may be used any time to answer calls, make calls, and check voicemail or e-mail.

Voicemail is also a valuable tool for the construction industry. If the party cannot be reached, the caller can leave a voicemail message. Most business telephones and cellular telephones have voicemail capabilities. In addition, most individuals have answering machines or voicemail services on their home or cellular telephones.

Unlike voicemail, pagers allow the caller to submit a message immediately, even when the recipient is away from a telephone. The message can be numeric, such as a telephone number, or it can be alphanumeric, such as "Call Linda on Project XYZ immediately at 555-1122."

Handheld communication devices allow construction supervisors to plan their calendars, schedule meetings, manage projects, and access their e-mail from remote locations. These computers are small enough to fit in the palm of your hand, yet powerful enough to hold years of information from various projects. Information can be transmitted electronically to others on the project team or transferred to a desktop computer.

SESSION 3

3.0.0 GENDER AND MINORITY ISSUES

To compensate for a low pool of skilled applicants, the construction industry has actively begun recruiting women and minorities.

In his book *Communication: The Essence of Labor Relations*, Richard Hutchinson reported that the number of women in the construction industry increased nearly 16 percent between 1995 and 1999. By the end of 1999, there were approximately 886,000 female construction workers versus 8,101,000 male construction workers in the industry.

The United States Census Bureau reported that there are approximately 24 million foreign-born residents who are becoming a greater proportion of the workforce in America. The Census Bureau predicts that non-whites will represent half of the United States population by 2050.

Some issues relating to gender and race must be addressed on the job site. These include different communication styles of men and women, language barriers associated with race, cultural differences, the possibility of sexual harassment, and the potential for cases of gender or racial discrimination.

3.1.0 Communication Styles of Men and Women

As more and more women move into construction, it becomes increasingly important that communication barriers between men and women are broken down and differences in behaviors are understood so that they can work together more effectively.

The Jamestown, New York Area Labor Management Committee (JALMC) offers the following explanations and tips.

1. *Women tend to ask more questions than men do.* Women who ask questions aren't ignorant; they are interested, and they want to learn. Men can support their learning by answering their questions patiently and adding a few questions of their own. That way, both genders are involved in the problem-solving and decision-making process.

2. Men tend to offer solutions before empathy; women tend to do the opposite. Both men and women should say what they want up-front—solutions or a sympathetic ear. That way, both genders will feel understood and supported.

3. Women are more likely to ask for help than men. Women should let men know in an easy-going way that they can and are willing to help. In addition, men should avoid doing a task for a woman when she asks for help; instead, they should show her how to do it so that she can do it alone the next time. That way, both men and women get the help that they need, and things are accomplished more quickly.

4. Men tend to communicate more competitively, and women tend to communicate more cooperatively. Women should learn to finish their statements, even if interrupted, and they should remember that interruptions are a man's way of asking questions and showing interest. Men should try to hear the woman out before interrupting. The result is a team or work unit that is both cooperative and competitive, a great combination.

5. *To establish trust with one another, women tend to self-disclose while men focus on reliability.* Women should try to keep all commitments with men, even small ones. In the event that they cannot be kept, women should attempt to renegotiate so that a positive track record of reliability is maintained. Men should try to understand that women who share their feelings are doing so because they feel that they can trust the other person, not because they want to appear unstable or weak. Recognizing these two different styles leads to less suspicion and hesitancy about working together, and contributes to a better flow of information.

3.2.0 Language Barriers

Language barriers are a real workplace challenge. The U.S. Census Bureau reported that the number of people who speak languages other than English increased from 23.1 million to 31.8 million between 1980 and 1990.

Spanish is now the most common non-English language spoken in the United States. As the make-up of the immigration population continues to change, the number of non-English speakers will rise dramatically, and the languages being spoken will also change.

According to an article in *USA Today*, the U.S. population will grow by about 130 million people by the year 2050. About one-third of those people will be from other countries. The article offers the following solutions based on what companies have done to overcome this challenge:

* Hire supervisors who speak the language of the non-English speaking workers

* Offer English classes either at the worksite or through school districts and community colleges

* Offer incentives for workers to learn English

* Ask employees who are parts of the problem to become parts of the solution

* Do not assume that the work groups common language must be English

* Allow work groups to pick the language that works best for them, except on occasions where a common language may be necessary

* Develop a policy statement about languages in the workplace

As our workforce becomes more diverse, communicating with people for whom English is a second language will be even more critical. The Knowledge Center's *Manager's Tool Kit* offers the following tips for communicating across language barriers:

* Be patient. If you expect too much, too quickly, you will likely confuse and frustrate the people involved.

* Don't get angry if people don't understand at first. Give them time to process the information in a way that they can comprehend.

* If people speak English poorly but understand reasonably well, ask them to demonstrate their understanding through actions rather than words.

- Avoid humor. Your jokes will likely be misunderstood and may even be misinterpreted as a joke at the worker's expense.

- Don't assume that people are unintelligent simply because they don't understand. Many immigrants doing menial jobs in their adopted countries were doctors, lawyers, and educators in their native countries.

- Speak slowly and clearly, and avoid the tendency to yell.

- Use face-to-face communication whenever possible. Over-the-phone communication is often more difficult when a language barrier is involved.

- Use pictures or drawings to get your point across, if necessary.

3.3.0 Cultural Differences

As workers from a multitude of backgrounds and cultures are brought together, there are bound to be differences and conflicts in the workplace.

To overcome cultural conflicts, the SHRM suggests the following:

1. Define the problem from both points of view. How does each person involved view the conflict? What does each person think is wrong? This involves moving beyond traditional thought processes to consider alternate ways of thinking.

2. Uncover cultural interpretations.What assumptions are being made based on cultural programming? By doing this, the supervisor may realize what motivated an employee to act in a particular manner. For instance, an employee may call in sick for an entire day to take his or her spouse to the doctor. In some cultures, such an action is based on responsibility and respect for family.

3. Create cultural synergy. Devise a solution that works for both parties involved. The purpose is to recognize and respect other's cultural values, and work out mutually acceptable alternatives.

3.4.0 Sexual Harassment

In today's business world, men and women are working side-by-side in careers of all kinds.

There are no longer "male occupations" or "female jobs"; therefore, men and women are expected to relate to each other in new ways that are uncharacteristic of the past.

As women make the transition into traditionally male industries, such as construction, the likelihood of sexual harassment increases. Sexual harassment is defined as unwelcome behavior of a sexual nature that makes someone feel uncomfortable in the workplace by focusing attention on their gender instead of on their professional qualifications. Sexual harassment can range from telling an offensive joke or hanging a poster of a swimsuit-clad man or woman in one's cubicle to making sexual comments or physical advances within a work environment.

Historically, sexual harassment was thought to be an act performed by men of power within an organization against women in subordinate positions with lesser power. However, as the number of sexual harassment cases have shown over the years, this is no longer the case.

Sexual harassment can occur in a variety of circumstances, including but not limited to the following:

- The victim as well as the harasser may be a woman or a man. The victim does not have to be of the opposite sex.

- The harasser can be the victim's supervisor, an agent of the employer, a supervisor in another area, a co-worker, or a non-employee.

- The victim does not have to be the person harassed but could be anyone affected by the offensive conduct.

- Unlawful sexual harassment may occur without economic injury to or discharge of the victim.

- The harasser's conduct must be unwelcome.

The Equal Employment Opportunity Commission (EEOC) enforces sexual harassment laws within industries. When investigating allegations of sexual harassment, the EEOC looks at the whole record including the circumstances and the context in which the alleged incidents occurred. A decision on the allegations is made from the facts on a case-by-case basis.

Prevention is the best tool to eliminate sexual harassment in the workplace. The EEOC encourages employers to take steps necessary to prevent sexual harassment from occurring. Employers should clearly communicate to employees that sexual harassment will not be tolerated. They can do so by developing a policy on sexual harassment, establishing an effective complaint or grievance process, and taking immediate and appropriate action when an employee complains.

3.5.0   Gender and Minority Discrimination

With the increase of women and minorities in the workforce, there is more room for gender and minority discrimination. Consequently, many business practices, including the way employees are treated, the organization's hiring and promotional practices, and the way people are compensated, are being analyzed for equity. More attention is being placed on fair recruitment, equal pay for equal work, and promotions for women and minorities in the workplace.

The EEOC requires that companies are equal opportunity employers. This means that organizations hire the best person for the job, without regard for race, sex, religion, age, etc. Once traditionally a male-dominated industry, construction companies are moving away from this notion and are actively recruiting and training women, younger workers, people from other cultures, and workers with disabilities to compensate for the shortage of skilled workers.

Despite the positive change in recruitment procedures, there are still inadequacies when it comes to pay and promotional activities associated with women and minorities. Individuals and federal agencies continue to win cases in which women and minorities have been discriminated against in the workplace, thus demonstrating that discrimination persists. The EEOC reported that between 1992 and 1997 two to three million dollars was awarded each year to those winning sexual discrimination cases.

To prevent discrimination cases, employers should develop valid job-related criteria for hiring, compensation, and promotion. These measures should be used consistently for every applicant interview, employee performance appraisal, etc. Therefore, all employees responsible for recruitment and selection, supervision, and evaluating job performance should be trained on how to use the job-related criteria legally and effectively.

REVIEW QUESTIONS ◆ CHAPTER ONE, SECTIONS 2.0.0–3.5.0

1. The construction industry should provide training for craftworkers and supervisors _____.

   a. to ensure that there are enough future workers

   b. since growth demands are anticipated to be greater than replacement needs

   c. in order to update the skills of older workers who are retiring at a later age than they previously did

   d. even though younger workers are now less likely to seek jobs in other areas than they were 10 years ago

2. Construction companies traditionally offer craftworker training _____.

   a. that a supervisor leads in a classroom setting

   b. that a craftworker leads in a classroom setting

   c. in a hands-on setting, where craftworkers learn from a co-worker or supervisor

   d. on a self-study basis to allow craftworkers to proceed at their own pace

3. One way to provide effective training is to _____.

   a. avoid giving negative feedback until trainees are more experienced in doing the task

   b. tailor the training to the career goals of trainees

   c. choose one kind of training method and stick with it for consistency's sake

   d. encourage trainees to listen, saving their questions for the end of the session

4. Among the benefits of new technology in the construction industry is _____.

   a. the Internet, because it can be used for exchanging and retrieving information

   b. the World Wide Web, because users can send messages worldwide

   c. the ability of e-mail to retrieve information on various topics that can then be distributed to multiple people

   d. e-commerce, because it offers Web pages that are solely informational in nature, which in turn helps bring publicity to a company

5. One way to prevent sexual harassment in the workplace is to _____.

   a. require employee training in which the potentially offensive subject of stereotypes is carefully avoided

   b. develop a consistent policy with appropriate consequences for engaging in sexual harassment

   c. communicate to workers that the victim of sexual harassment is the one who is being directly harassed, not those affected in a more indirect way

   d. educate workers to recognize sexual harassment for what it is—unwelcome conduct by the opposite sex

6. Employers can minimize all types of workplace discrimination by doing the following *except* _____.

   a. investigating those complaints that are serious

   b. examining if there are large differences in number between the work force and qualified workers

   c. hiring based on a consistent list of job-related requirements

   d. providing sexual harassment training

SECTION 4

4.0.0 CONSTRUCTION PROJECTS

There are four major categories of construction projects: residential, commercial and institutional, industrial, and civil. Each fills the needs of the industry in its own way.

4.1.0 Major Categories of Construction Projects

In all four construction categories, there are several common threads that unite the industry:

- Construction helps satisfy the basic need for shelter.
- Special skills and professions are required.
- Material and technology continue to improve.
- The building structure reflects the changing needs of society.

4.1.1 Residential Construction

Residential construction refers to the building of single and multifamily homes, as well as the repair or remodeling of existing homes. Many construction trade workers begin their careers in this field.

4.1.2 Commercial and Institutional Construction

Commercial and *institutional* construction includes building and modernizing stores, hospitals, schools, apartment houses, warehouses, restaurants, and office buildings. These projects may range from small-scale jobs, involving only a minimum of time and cost, to large-scale jobs with contracts of several million dollars, which last many years.

Field personnel with many specialties are involved in commercial and industrial construction. Supervisors must be more involved in the activities than in residential construction. In addition, commercial and institutional construction projects usually involve the use of materials and systems not found in residential construction. Such materials include structural steel, pre-cast concrete, sprinkler systems, escalators, and elevators.

4.1.3 Industrial Construction

Industrial construction refers to building new manufacturing plants and modernizing and expanding existing plants. Because manufacturing itself is so diverse, industrial construction projects vary widely. They range from building an addition to a small electronics assembly plant to the construction of super-sized nuclear power plants.

4.1.4 Civil Construction

Civil construction refers to large projects needing civil engineering design. Examples include the construction of roads, bridges, runways, utilities, and dams.

REVIEW QUESTIONS ◆ CHAPTER ONE, SECTIONS 4.0.0–4.1.4

1. One of the four major categories of construction projects is _____.

   a. civil, which involves such projects as road and bridge construction

   b. residential, which involves any sort of dwelling such as homes as well as apartments

   c. commercial and institutional, which use materials similar to those in residential construction but on a larger scale

   d. institutional and industrial, which involve schools, offices, and factories

2. Many construction trade workers begin their careers in _____ construction.

   a. commercial and institutional

   b. industrial

   c. civil

   d. residential

3. Materials such as structural steel, sprinkler systems, and elevators are used in _____ construction.

   a. residential

   b. commercial and institutional

   c. civil

   d. industrial

SECTION 5

5.0.0. THE CONSTRUCTION ORGANIZATION

An *organization* is concerned with the relationships among the people within the company. The supervisor needs to be aware of two types of organizations. These are *formal* organizations and *informal* organizations.

A formal organization exists when the activities of the people within the work group are coordinated toward the attainment of a goal. An example of a formal organization is a work crew consisting of four carpenters and two laborers led by a supervisor all working together on a job.

A formal organization is typically documented on an organizational chart, which outlines all of the positions that make up an organization and how those positions are related. Some organizational charts even depict the people within each position and the person to whom they report as well as the people that the person supervises, which is also known as *span of control*. An efficient span of control consists of a manager who directly supervises no more than six people.

An informal organization can be described as a group of individuals having a common goal, but for different reasons. An example of this would be a plane full of passengers flying to the same city, each with a different reason for doing so, and dispersing once the city is reached.

An informal organization allows for communication among its members so they can perform as a group. It also establishes patterns of behavior that help them to work as a group. One pattern may be a specific manner of dressing, speaking alike, or developing conformity.

Both types of organizations establish the foundation for how communications flow. The formal structure is the means used to delegate authority and responsibility and to exchange information. The informal structure is used to exchange information. The "grapevine" is an example of an informal organization.

Members in an organization perform best when each member:

- Understands the job and how it will be done
- Communicates effectively with others in the group
- Knows his or her role in the group
- Knows who has the authority and responsibility

5.1.0 Division of Responsibility

The conduct of a construction business involves certain functions. In a small organization, responsibilities may be divided between one or two people. However, in a larger organization with many different and complex activities, responsibilities may be grouped into similar activity groups, and the responsibility for each group assigned to department managers. In either case, the following major departments exist in construction companies:

Executive: This office represents top management. It is responsible for the success of the company through short-range and long-range planning.

Accounting: This office is responsible for all record-keeping and financial transactions, including payroll, taxes, insurance, and audits.

Contract Administration: This office prepares and executes contractual documents with owners, sub-contractors, and suppliers.

Purchasing: This office obtains material prices and then issues purchase orders. The purchasing office also obtains rental and leasing rates on equipment and tools.

Estimating: This office is responsible for recording the quantity of material on the jobs, the "takeoff", pricing labor and material, analyzing sub-contractors' bids, and bidding on projects.

Construction Operation: This office plans, controls, and supervises all construction field activities.

Other divisions of responsibility a company may create involve architectural and engineering design functions. These divisions usually become separate departments.

5.2.0 Authority and Responsibility

As the organization grows, the manager must ask others to perform many duties so that the manager can concentrate on managing. Managers typically assign activities to their subordinates or workers through a process called *delegation*.

When delegating activities, the supervisor assigns others the responsibility to perform the designated tasks. Delegation implies responsibility, which refers to the obligation of an employee to perform the duties.

Along with responsibility comes *authority*. Authority is the power to act or make decisions in carrying out an assignment. The kind and amount of authority a supervisor or worker has depends on the company for which he or she works.

Authority and responsibility must be balanced so employees can carry out their tasks. In addition, delegation of sufficient authority is needed to make an employee accountable to the supervisor for the results. Accountability is the act of holding an employee responsible for completing the assigned activities. Even though authority is delegated, the final responsibility always rests with the supervisor.

5.3.0 Job Descriptions

Many companies furnish each employee a written job description, which explains the job in detail. Job descriptions set a standard for the employee, make judging performance easier, clarify the tasks each person should handle, and simplify the training of new employees.

Each new employee should understand all the duties and responsibilities of the job after reviewing the job description. Thus, the time it takes for the employee to make the transition from being a new and uninformed employee to a more experienced member of a production team is shortened.

A job description need not be long, but it should be detailed enough to ensure there is no misunderstanding of the duties and responsibilities of the position. The job description should contain all the information necessary to evaluate the employee's performance.

A job description should contain the following:

- Job title
- General description of the position
- Specific duties and responsibilities
- The supervisor to whom the position reports
- Other requirements, such as required tools/equipment for the job

A sample job description is shown in *Figure 1*.

Position:

Front-line Construction Supervisor – Field

General Summary:

First line of supervision on a construction crew installing underground and underwater cathodic protection systems.

Reports To:

Field Superintendent

Duties and Responsibilities:

- Oversee crew
- Provide instruction and training in construction tasks as needed
- Make sure proper materials and tools are on the site to accomplish tasks
- Keep project on schedule
- Enforce safety regulations

Knowledge, Skills, And Experience Required:

- Extensive travel throughout the Eastern United States, home base in Atlanta
- Ability to operate a backhoe and trencher
- Valid commercial drivers license with no DUI violations
- Ability to work under deadlines with the knowledge and ability to foresee problem areas and develop a plan of action to solve the situation

---

Figure 1 • Job Description

5.4.0 Policies and Procedures

Most construction companies have formal policies and procedures established to help the supervisor carry out his or her duties. A policy is a general statement establishing guidelines for a specific activity. Examples include policies on vacations, rest breaks, workplace safety, or checking out power hand tools.

Procedures are the ways that policies are carried out. For example, a procedure written to implement a policy on workplace safety would include guidelines for reporting accidents and general safety procedures that all employees are expected to follow.

A sample of a policy and procedure is shown in *Figure 2*.

WORKPLACE SAFETY POLICY:

Your safety is the constant concern of this company. Every precaution has been taken to provide a safe workplace.

Common sense and personal interest in safety are still the greatest guarantees of your safety at work, on the road, and at home. We take your safety seriously and any willful or habitual violation of safety rules will be considered cause for dismissal. We are sincerely concerned for the health and well being of each member of the team.

Workplace Safety Procedure:

To ensure your safety, and that of your co-workers, please observe and obey the following rules and guidelines:

- Observe and practice the safety procedures established for the job.
- In case of sickness or injury, no matter how slight, report at once to your supervisor. In no case should an employee treat his own or someone else's injuries or attempt to remove foreign particles from the eye.
- In case of injury resulting in possible fracture to legs, back, or neck, any accident resulting in an unconscious condition, or a severe head injury, the employee is not to be moved until medical attention has been given by authorized personnel.
- Do not wear loose clothing or jewelry around machinery. It may catch on moving equipment and cause a serious injury.
- Never distract the attention of another employee, as you might cause him or her to be injured. If necessary to get the attention of another employee, wait until it can be done safely.
- Where required, you must wear protective equipment, such as goggles, safety glasses, masks, gloves, hairnets, etc.
- Safety equipment such as restraints, pull backs, and two-hand devices are designed for your protection. Be sure such equipment is adjusted for you.
- Pile materials, skids, bins, boxes, or other equipment so as not to block aisles, exits, fire-fighting equipment, electric lighting or power panel, valves, etc. FIRE DOORS AND AISLES MUST BE KEPT CLEAR.
- Keep your work area clean.
- Use compressed air only for the job for which it is intended. Do not clean your clothes with it and do not play with it.

Figure 2 • Safety Policy and Procedure

REVIEW QUESTIONS ◆ CHAPTER ONE, SECTIONS 5.0.0–5.4.0

1. Members tend to function best within an organization when they _____.
   a. are allowed to select their own style of clothing for each project
   b. know their role
   c. do not disagree with the statements of other workers or supervisors
   d. have authority

2. A formal organization is defined as _____.
   a. a structure used for exchanging information, as well as for delegating responsibility
   b. an entity that establishes standards such as a dress code to encourage the idea of teamwork
   c. a group of individuals who have their own personal motivations but who share the same goal
   d. a structure with basic hierarchies in place

3. A formal organization uses an organizational chart to _____.
   a. depict all companies with whom it conducts business
   b. show all customers with whom it conducts business
   c. show the relationships among the existing positions in the generic sense, typically excluding the actual people within the position
   d. show the relationships among the existing positions; may even include the actual people in the position

4. Authority is defined as _____.
   a. making decisions without ever having to discuss things with others
   b. giving an employee a particular task to perform
   c. making an employee responsible for the completion and results of a particular duty
   d. having the power to perform an action or make decisions, such as promoting someone

5. A good job description should include _____.
   a. a complete organization chart
   b. any information needed to judge the job performance of an individual
   c. a long, detailed description of the position
   d. the company's sexual harassment policy

6. Job descriptions are used in order to _____.
   a. specify to job candidates whether a company uses a formal or informal organizational structure
   b. give details to potential employees about how to perform specific duties
   c. give information regarding the responsibilities of the potential job candidate
   d. inform the potential candidate about the company's benefits

7. The purpose of a policy is to _____.
   a. implement company guidelines regarding a particular activity
   b. specify what tools and equipment are required for a job
   c. list all information necessary to judge an employee's performance
   d. inform employees about the future plans of the company

8. The following are examples of a procedure *except* _____.
   a. rules for taking time off
   b. guidelines for receiving additional training
   c. general comments about the company's intolerance for gender discrimination
   d. ways to avoid injury when welding

CHAPTER TWO

Leadership Skills

OBJECTIVES

Upon completion of this chapter, you should be able to:

1. Explain the role of a crew leader.

2. List the characteristics of effective leaders.

3. Be able to discuss the importance of ethics in a supervisor's role.

4. Identify the three styles of leadership.

5. Describe the forms of communication.

6. Explain the four parts of verbal communication.

7. Demonstrate the importance of active listening.

8. Illustrate how to overcome the barriers to communication.

9. List some ways that supervisors can motivate their employees.

10. Explain the importance of delegating and implementing policies and procedures.

11. Differentiate between problem solving and decision making.

SECTION 1

1.0.0 INTRODUCTION TO SUPERVISION

For the purpose of this program, it is important to define some of the positions to which we will be referring. The term *craftworker* refers to a person who performs the work of his or her trade(s). The *crew leader* is a person who supervises one or more craftworkers on a crew. A *superintendent* is essentially a second-line supervisor who supervises one or more crew leaders or front-line supervisors. Finally, a *project manager* is responsible for managing the construction of one or more construction projects. In this training, we will concentrate primarily on the role of the crew leader.

The organizational chart in *Figure 3* on the following page depicts how each of the four positions is related in the chain of command.

Craftworkers and crew leaders differ in that the crew leader manages the activities that the skilled craftworkers on the crews actually perform. In order to manage a crew of craftworkers, a crew leader must have first-hand knowledge and experience in the activities being performed. In addition, he or she must be able to act directly in organizing and directing the activities of the various crew members.

This chapter will discuss the importance of developing effective leadership skills as a new supervisor as well as provide some tips for doing so. Effective ways to communicate with all levels of employees, build teams, motivate crew members, make decisions, and resolve problems will be covered in depth.

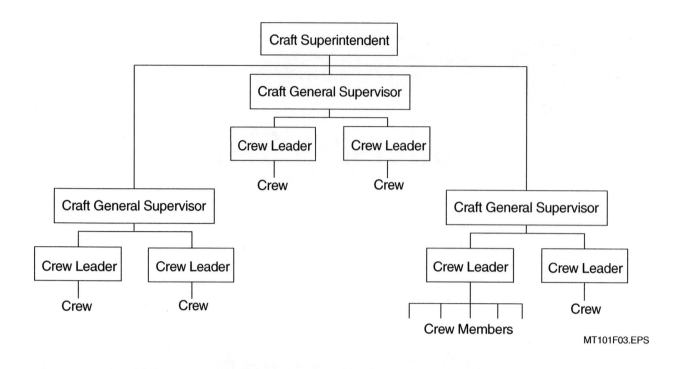

Figure 3 • Project Organization Chart

SECTION 2

2.0.0   THE SHIFT IN WORK ACTIVITIES

The crew leader is generally selected and promoted from a construction work crew. The selection will frequently be based on that person's ability to accomplish tasks, to get along with others, to meet schedules, and to stay within the budget. The crew leader must lead the team to work safely and provide a quality product.

Making the transition from a craftworker to a crew leader can be difficult, especially when the new supervisor is in charge of supervising a group of peers. Crew leaders are no longer responsible for their work alone; rather, they are accountable for the work of an entire crew of people with varying skill levels and abilities, a multitude of personalities and work styles, and different cultural and educational backgrounds.

Supervisors must learn to put their personal relationships aside and work for the common goals of the team.

As an employee moves from a craftworker position to the role of a crew leader, he or she will find that more hours will be spent supervising the work of others than actually performing the technical skill for which he or she has been trained. *Figure 4* represents the percentage of time craftworkers, crew leaders, superintendents, and project managers spend on technical and supervisory work as their management responsibilities increase.

The success of the new crew leader is directly related to the ability to make the transition from crew member into a leadership role.

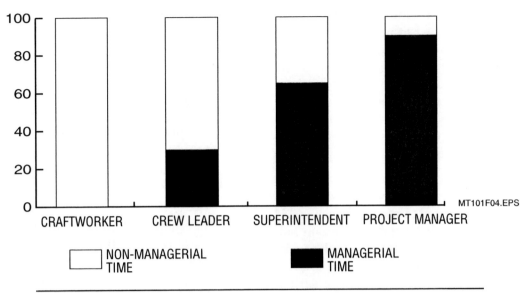

Figure 4 • Percentage of Time Spent on Technical and Supervisory Work

SECTION 3

3.0.0 BECOMING A LEADER

It is essential that a supervisor have leadership skills to be successful. Therefore, one of the primary goals of a person who wants to become a supervisor should be to develop strong leadership skills and learn to use them effectively.

The expression a *natural born leader* is pretty common. Although there is no overwhelming evidence for this, it is true that some people may have inherited leadership characteristics or developed traits that motivate others to follow and perform. Research shows that people who possess such talents are likely to succeed as leaders.

3.1.0 Characteristics of Leaders

Leadership traits are similar to the skills that a supervisor needs to be effective. Although the characteristics of leadership are many, there are some definite commonalities among effective leaders.

First and foremost, effective leaders lead by example. In other words, they work and live by the standards that they establish for their crew members or followers, making sure that they "practice what they preach".

Next, effective leaders tend to have a high level of drive, determination, persistence, or a "stick-to-it" attitude. In the face of a challenging situation, leaders seek involvement and drive through adversity to achieve their goal. When faced with obstacles, effective leaders don't get discouraged; instead, they identify the potential problems, make plans to overcome them, and work toward achieving the intended goal. In the event of failure, effective leaders learn from their mistakes and apply that knowledge to their future leadership attempts.

Third, effective leaders are typically effective communicators. Accomplishing this may require that the leader overcome issues such as language barriers, gender biases, or differences in personalities to ensure that each follower or member of the crew understands the established goals of the project.

Effective leaders have the ability to motivate their followers or crew members to work to their full potential and become effective members of the team. They try to develop their crew members' skills and encourage them to improve and learn as a means to contribute more to the team effort. Effective leaders tend to demand 100% from themselves and their teams, so they work hard to provide the skills and leadership necessary to do so.

In addition, effective leaders are organized planners. They know what needs to be accomplished, and they use their resources to make it happen. Because they cannot do it alone, leaders enlist the help of their team members to share in the workload. Effective leaders delegate work to their crew members, and they implement company policies and procedures to ensure that the work is completed effectively, efficiently, and safely.

Finally, effective leaders have the self-confidence that allows them to make decisions and solve problems. If they are going to accomplish their goals, leaders must learn to take risks when appropriate. Leaders must be able to absorb information, assess courses of action, understand the risks, make decisions, and assume the responsibility for them.

3.1.1   Leadership Traits

There are many other traits of effective leaders. In fact, the list is very long. Some other major characteristics of leaders include:

- The ability to plan and organize
- Loyalty
- Fairness
- Enthusiasm
- The ability to teach and learn from others
- Initiative
- Salesmanship
- Good communication skills

3.1.2   Expected Leadership Behavior

Followers have expectations of their leaders. They look to their leaders to:

- Set the example
- Suggest and direct
- Influence their actions
- Communicate effectively
- Make decisions and assume responsibility
- Be a loyal member of the group
- Abide by company policies and procedures

3.2.0   Functions of a Leader

The functions of a leader will vary with the environment, the group being led, and the tasks to be performed. However, there are certain functions common to all situations which the leader will be called upon to perform. Some of the major functions are:

- Organize, plan, staff, direct, and control
- Empower group members to make decisions and take responsibility for their work
- Maintain a cohesive group by resolving tensions and differences among its members and between the group and those outside the group
- Ensure that all group members understand and abide by company policies and procedures
- Accept responsibility for the success and failures of the group's performance
- Represent the group to others
- Be sensitive to the differences of a diverse workforce

3.3.0   Leadership Styles

There are three broad categories of leadership style that supervisors can adopt. At one extreme is the autocratic or dictator style of leadership, where the supervisor makes all of the decisions independently, without seeking the opinions of crew members. At the other extreme is the hands-off leadership style, where the supervisor lets the employees make all of the decisions. In the middle is the democratic style, where the supervisor seeks crew members' opinions and makes the appropriate decisions based on their input.

The following are some characteristics of each of the three leadership styles:

Autocratic leaders:

- Expect their crew members to work
- Seldom seek advice from their crew members
- Insist on solving problems alone
- Seldom permit crew members to assist each other
- Praise and criticize on a personal basis
- Have no sincere interest in improving methods of operation or in increasing production

Democratic leaders:
- Discuss problems with their crew members
- Listen
- Explain and instruct
- Give crew members a feeling of accomplishment by commending them when they do a job well
- Are friendly and available to discuss personal and job-related problems

Hands-off leaders:
- Believe no supervision is best
- Rarely give orders
- Worry about whether they are liked by their crew members

Effective leadership takes many forms. The correct style for a particular situation or operation depends on the nature of the crew as well as the work it has to accomplish. For example, if the crew does not have enough experience for the job ahead, then a direct, autocratic style may be appropriate. The autocratic style of leadership is also effective when jobs involve repetitive operations that require little decision-making.

However, if a worker's attitude is an issue, then democratic action is required. In this case, providing the missing motivational factors may increase performance and result in the improvement of the worker's attitude. The democratic style of leadership is also used when the work is of a creative nature because brainstorming and exchanging ideas with the crew members can be beneficial.

Hands-off leadership is not a very effective leadership style. In these cases, either the leadership becomes the responsibility of a crew member who takes the initiative to assume the role, or chaos results due to lack of formal leadership from the supervisor.

3.3.1 Selecting a Leadership Style

In selecting the most appropriate style of leadership, the supervisor should consider the power base from which to work. There are three elements that make up a power base:

1. Authority
2. Expertise, experience, and knowledge
3. Respect, attitude, and personality

First, the company must give a supervisor sufficient authority to do the job. This authority must be commensurate with responsibility, and it must be made known to crew members when they are hired so that they understand who is in charge.

Next, a supervisor must have an expert knowledge of the activities to be supervised in order to be effective. This is important because the crew members need to know that they have someone to turn to when they have a question or a problem, when they need some guidance, or when modifications or changes are warranted by the job.

Finally, respect is probably the most useful element of power. This usually derives from being fair to employees, by listening to their complaints and suggestions, and by utilizing incentives and rewards appropriately to motivate crew members. In addition, supervisors who have a positive attitude and a favorable personality tend to gain the respect of their crew members as well as their peers. Along with respect comes a positive attitude from the crew members.

3.4.0 Ethics in Leadership

The construction industry demands the highest standards of ethical conduct. Every day the crew leader has to make decisions which may have ethical implications. When an unethical decision is made, it not only hurts the supervisor but also other workers, peers, and the company for which he or she works.

There are three basic types of ethics:

1. Business, or legal
2. Professional, or balanced
3. Situational

Business, or legal, ethics is adhering to all laws and regulations related to the issue. Professional, or balanced, ethics is carrying out all activities in such a manner as to be fair to everyone concerned. Situational ethics pertains to specific activities or events that may initially appear to be a "gray area." For example, you may ask yourself, "How will I feel about myself if my actions are published in the newspaper or if I have to justify my actions to my family, friends, and colleagues?"

Let's consider the example of professional football. The National Football League has rules by which the game is played. If someone breaks one of the rules, he is acting in an unethical manner from a business or legal sense. It is an accepted standard among professional football players that when they tackle another player it is to take him out of the play and not to purposely injure him. If one were to tackle someone with the motivation of causing an injury, this would be considered unethical.

If you were running down the side line after catching a pass to score a touch down, and in your attempt slightly stepped out of bounds (without anyone calling the ball dead at that point), and you did not voluntarily acknowledge that you did, you would be unethical from a situational standpoint. How would you feel the next day if a newspaper printed a picture showing your foot out of bounds and insinuating that your character is questionable because you did not acknowledge breaking the rules?

There are going to be many times when the supervisor will be put into a situation where he or she will need to assess the ethical consequences of an impending decision. For instance, should a supervisor continue to keep one of his or her crew working who has broken into a cold sweat due to overheated working conditions just because the superintendent says the activity is behind schedule? Or should a supervisor, who is the only one aware that the reinforcing steel placed by his or her crew was done incorrectly, correct the situation before the concrete is placed in the form? If a supervisor is ever asked to carry through on an unethical decision, it is up to him or her to inform the supervisor of the unethical nature of the issue, and if still requested to follow through, refuse to act.

SECTION 4

4.0.0 COMMUNICATION

Successful supervisors learn to communicate effectively with people at all levels of the organization. In doing so, they develop an understanding of human behavior and acquire communication skills that enable them to understand and influence others.

There are many definitions for *communication*. One is that communication is the act of accurately and effectively conveying or transmitting facts, feelings, and opinions to another person. Simply stated, communication is the method of exchanging information and ideas.

Just as there are many definitions of communication, it also comes in many forms. Verbal, non-verbal, and written are the types of communication. Each of these forms of communication will be discussed in this section.

Research shows that the typical supervisor spends about 80 percent of his or her day communicating through writing, speaking, listening, or using "body language." Of that time, studies suggest that approximately 20 percent of communication is written, and 80 percent involves speaking or listening.

4.1.0 Verbal Communication

Verbal communication refers to the spoken words exchanged between two or more people when communicating. It can be done face-to-face or over the telephone.

Verbal communication consists of four distinct parts:

1. Sender
2. Message
3. Receiver
4. Feedback

The diagram in *Figure 5* depicts the relationship of these four parts within the communication process.

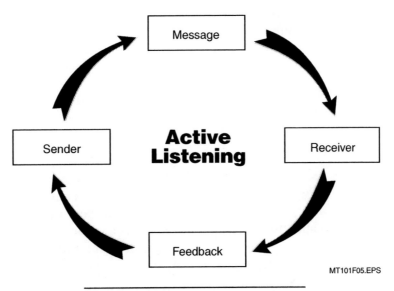

Figure 5 • Communication Process

4.1.1 The Sender

The *sender* is the person that creates the message to be communicated. In verbal communication, the sender actually says the message aloud to the person(s) for whom it is intended.

The sender must be sure to speak in a clear and concise manner that can be easily understood by others. This is not an easy task; it takes practice.

Some basic tips to consider:

- Avoid talking with anything in your mouth (food, gum, pencils, etc.).

- Find an appropriate rate of speech; don't talk too quickly or too slowly. In extreme cases, people tend to focus on your rate of speech instead of on what you are saying.

- Be aware of your tone. Remember, it's not so much what you say, but how you say it.

- Enunciate to prevent misunderstandings (many letters, such as T, D, B, and E sound similar).

- Don't talk in a monotone; rather, put some enthusiasm and feeling in your voice.

4.1.2 The Message

The *message* is what the sender is attempting to communicate to the audience. A message can be a set of directions, an opinion, or a feeling. Whatever its function, a message is an idea or fact that the sender wants the audience to know.

Before speaking, the sender should determine what it is he or she wants to communicate. The sender should then organize what to say, ensuring that the message is logical and complete. Taking the time to clarify your thoughts prevents rambling, not getting the message across effectively, or confusing the audience. It also permits the sender to get to the point quickly.

In delivering the message, the sender should assess the audience. It is important not to talk down to your audience. Remember that everyone, whether in a senior or junior position, deserves respect and courtesy. Therefore, the sender should use words and phrases that the audience can understand and avoid technical language or slang. In addition, the sender should use short sentences, which give the audience time to understand and digest one point or fact at a time.

4.1.3 The Receiver

The *receiver* is the person to whom the message is communicated. For the communication process to be successful, it is important that the receiver understands the message as the sender intended. Therefore, the receiver must listen to what is being said.

The first step to becoming a good listener involves realizing that there are many barriers that get in the way of our listening, particularly on a busy construction job site.

Some barriers to effective listening are:

- Noise, visitors, telephone, or other distractions
- Preoccupation, being under pressure, or daydreaming
- Reacting emotionally to what is being communicated
- Thinking about how to respond instead of listening
- Giving an answer before the message is complete
- Personal biases to the sender's communication style
- Finishing the sender's sentence

Some tips for overcoming these barriers are:

- Take steps to minimize or remove distractions; learn to tune-out your surroundings
- Listen for key points
- Take notes
- Try not to take things personally
- Allow yourself time to process your thoughts before responding
- Let the sender communicate the message without interruption
- Be aware of your personal biases, and try to stay open-minded

There are many ways for a receiver to show that he or she is actively listening to what is being said. This can even be accomplished without saying a word. Examples include maintaining eye contact, nodding your head, and taking notes. It may also be accomplished through feedback.

4.1.4 Feedback

Feedback refers to the communication that occurs after the message has been sent by the sender and received by the receiver. It involves the receiver responding to the message.

Feedback is a very important part of the communication process because it allows the receiver to communicate how he or she interpreted the message, and it permits the sender to ensure that the message was understood as intended. In other words, feedback is a checkpoint to make sure that the receiver and the sender are "on the same page".

The receiver can use the opportunity of providing feedback to paraphrase back what was heard. When paraphrasing what you heard, it is best to use your own words. That way, you can show the sender that you interpreted the message correctly and could explain it to others if needed.

In addition, the receiver can clarify the meaning of the message and request additional information when providing feedback. This is generally accomplished by asking questions.

4.2.0 Non-Verbal Communication

Unlike verbal or written communication, *non-verbal* communication does not involve the spoken word or the written word. Rather, non-verbal communication refers to things that you can actually see when communicating with others. Examples include facial expressions, body movements, hand gestures, and eye contact.

Non-verbal communication can provide an external signal of an individual's inner emotions. It occurs simultaneously with verbal communication; often, the sender of the non-verbal communication is not even aware of it.

Because it can be physically observed, non-verbal communication is just as important as the words used in conveying the message. Often, people are influenced more by non-verbal signals than by spoken words. Therefore, it is important that you be conscious of the non-verbal cues that you send because you don't want the receiver to interpret your message incorrectly based on your posture or an expression on your face. After all, these things may have nothing to do with the current communication exchange; instead, they may be carrying over from something else going on in your day.

4.3.0 Written or Visual Communication

Some communication will have to be written or visual. *Written* or *visual* communication refers to communication that is documented on paper or transmitted electronically via the computer using words or visuals.

Many messages on a job have to be communicated in text form. Examples include weekly reports, requests for changes, purchase orders, and correspondence on a specific subject. These items are written because they must be recorded for business and historical purposes. In addition, some communication on the job will have to be visual. Items that are difficult to explain verbally or by the written word can best be explained through diagrams or graphics. Examples include the plans or drawings used on a job.

When writing or creating a visual message, it is best to assess the reader or the audience before beginning. It is essential that the reader can read the message and understand the content; otherwise, the communication process will be unsuccessful. Therefore, the writer should consider the actual meaning of words or diagrams and how others might interpret them. In addition, the writer should make sure that all handwriting is legible if the message is being handwritten.

Here are some basic tips for writing:

- Avoid emotion-packed words or phrases
- Be positive whenever possible
- Avoid using technical language or jargon
- Stick to the facts
- Provide an adequate level of detail
- Present the information in a logical manner
- Avoid making judgements, unless asked to do so
- Proofread your work; check for spelling and grammatical errors
- Make sure that the document is legible
- Be prepared to provide a verbal or visual explanation, if needed

Here are some basic tips for creating visuals:

- Provide an adequate level of detail
- Ensure that the diagram is large enough to be seen
- Avoid creating complex visuals; simplicity is better
- Present the information in a logical order
- Be prepared to provide a written or verbal explanation of the visual, if needed

4.4.0 Communication Issues

It is important to note that each person communicates a little differently; that is what makes us unique as individuals. And as the diversity of the workforce changes, communication will become even more challenging. We must consider that our audience consists of individuals from different ethnic groups, cultural backgrounds, educational levels, or economic status groups. Therefore, we must learn to assess our audience to determine how to effectively communicate with each individual.

The key to effective communication is to acknowledge that people are different and be able to adjust your communication style to meet the needs of your audience or the person on the receiving end of your message. This involves relaying your message in the simplest way possible, avoiding the use of words that people may find confusing. Be aware of how you use technical language, slang, jargon, and words that have multiple meanings. Present the information in a clear, concise manner. Avoid rambling and always speak clearly, using good grammar.

In addition, you have to be able to communicate your message in multiple ways or adjust your level of detail or terminology to ensure that everyone understands your meaning as intended. For instance, you may have to draw a map for a visual person who cannot comprehend directions in a verbal or written form, or overcome language barriers on the job site by using graphics or visual aids to relay your message.

Figure 6 shows how to tailor your message to your audience.

VERBAL INSTRUCTIONS Experienced Crew	VERBAL INSTRUCTIONS Newer Crew	WRITTEN INSTRUCTIONS	DIAGRAM/MAP
"Please drive to the supply shop to pick up our order."	"Please drive to the supply shop. Turn right here and left at Route 1. It's at 75th Street and Route 1. Tell them the company name and that you're there to pick up our order."	1. Turn right at exit. 2. Drive 2 miles to Route 1. Turn LEFT. 3. Drive 1 mile (pass the tire shop) to 75th Street. 4. Look for supply store on right. . . .	

Different people learn in different ways. Be sure to communicate so you can be understood.

MT101F06.EPS

Figure 6 • Tailor Your Message

PARTICIPANT EXCERCISE A

1. Read the following verbal conversations, and identify any problems:

Conversation I:

Judy: Hey Roger…

Roger: What's up?

Judy: Has the site been prepared for the job trailer yet?

Roger: Job trailer?

Judy: The job trailer – it's coming in today. What time will the job site be prepared?

Roger: The trailer will be here about 1:00 PM?

Judy: The job site! What time will the job site be prepared?

Conversation II:

John: Hey, Mike. I need your help.

Mike: What is it?

John: You and Joey go over and help Al's crew finish laying out the site.

Mike: Why me? I can't work with Joey. He can't understand a word I say.

John: Al's crew needs some help, and you and Joey are most qualified to do the job.

Mike: I told you, I can't work with Joey.

Conversation III:

Ed: Hey, Jill

Jill: Sir?

Ed: Have you received the latest DOL, EEO requirement to be sure the OFCP administrator finds our records up to date when he reviews them in August?

Jill: DOL, EEO, and OFCP?

Ed: Oh, and don't forget the MSHA, OSHA, and EPA reports are due this afternoon.

Jill: MSHA, OSHA, and EPA?

_____

_____

_____

_____

_____

_____

Conversation IV:

Susan: Hey, Bob, would you do me a favor?

Bob: Okay, Sue. What is it?

Susan: I was reading the concrete inspection report and found the concrete in Bays 4A, 3B, 6C, and 5D didn't meet the 3000 psi strength requirements. Also, the concrete inspector on the job told me the two batches that came in today had to be refused because they didn't meet the slump requirements as noted on page 16 of the spec. I need to know if any placement problems happened on those bays, how long the ready mix trucks were waiting today, and what do we plan to do to stop these problems in the future.

_____

_____

_____

_____

_____

2. Read the following written memos, and identify any problems:

Memo I:

Let's start with the transformer vault $285.00 due. For what you ask? Answer practically nothing I admit, but here is the story. Paul the superintendent decided it was not the way good ole Comm Ed wanted it, we took out the ladder and part of the grading (as Paul instructed us to do) we brought it back here to change it. When Comm Ed the architect or Doe found out that everything would still work the way it was, Paul instructed us to reinstall the work. That is the whole story there is please add the $285.00 to my next payout.

_____

_____

_____

_____

_____

_____

Memo II:

Let's take rooms C 307-C-312 and C-313 we made the light track supports and took them to the job to erect them when we tried to put them in we found direct work in the way, my men spent all day trying to find out what do so ask your Superintendent (Frank) he will verify seven hours pay for these men as he went back and forth while my men waited. Now the Architect has changed the system of hanging and has the gall to say that he has made my work easier, I can't see how. Anyway we want an extra two (2) men seven (7) hours for April 21 at $55.00 per hour or $385.00 on April 28th Doe Reference 197 finally resolved this problem. We will have no aditional charges on Doe Reference 197, please note.

_____

_____

_____

_____

_____

SECTION 5

5.0.0 MOTIVATION

The ability to motivate others is a key leadership skill that effective supervisors must possess. *Motivation* refers to behavior set into action because of a need felt by the individual to perform. It's also the word we use to describe the amount of effort that a person is willing to put forth to accomplish something. For example, a crew member who skips breaks and lunch in an effort to complete a job on time is thought to be highly motivated, but a crew member who does the bare minimum or just enough to keep his or her job is considered unmotivated.

Employee motivation has dimension because it can be measured. Examples of how motivation can be measured include determining the level of absenteeism, the percentage of employee turnover, and the number of complaints, as well as the quality and quantity of work produced.

5.1.0 Employee Motivators

Different things motivate different people in different ways. Consequently, there is no "one-size-fits-all" approach to motivating crew members. It is important to recognize that what motivates one crew member may not motivate another. In addition, what works to motivate a crew member once may not motivate that same person again in the future.

Frequently, the needs that motivate individuals are the same as those that create job satisfaction. They include:

- Recognition and praise
- Accomplishment
- Opportunity for advancement
- Job importance
- Change
- Personal growth
- Rewards

A supervisor's ability to satisfy these needs increases the likelihood of high morale within a crew. *Morale* refers to individuals' attitudes toward the tasks that they are expected to perform. High morale, in turn, means that employees will be motivated to work hard, and they will have a positive attitude about coming to work and doing their jobs.

5.1.1 Recognition and Praise

Recognition and *praise* refer to the need to have good work appreciated, applauded, and acknowledged by others. This can be accomplished by simply thanking employees for helping out on a project, or it can entail more formal praise, such as an award for "Employee of the Month".

Some tips for giving recognition and praise:

- Be available on the job site so that you have the opportunity to witness good work
- Know good work, and praise it when you see it
- Look for good work and look for ways to praise it
- Give recognition and praise only when truly deserved; otherwise, it will lose its meaning
- Acknowledge satisfactory performance, and encourage improvement by showing that you have confidence in the ability of your crew members to do above-average work

5.1.2 Accomplishment

Accomplishment refers to a worker's need to set challenging goals and achieve them. There is nothing quite like the feeling of achieving a goal, particularly a goal one never expected to accomplish in the first place.

Supervisors can help their crew members attain a sense of accomplishment by encouraging them to develop performance plans, such as goals for the year. In addition, supervisors can provide the support and the tools (such as training and coaching) necessary to help their crew members achieve these goals.

5.1.3 Opportunity for Advancement

Opportunity for advancement refers to an employee's need to gain additional responsibility and develop new skills and abilities. It is important that some employees know that they are not stuck in "dead-end" jobs. Rather, they want a chance to grow with the company and to be promoted in recognition of excelling in their work.

Effective leaders encourage their crew members to work to their full potentials. In addition, they share information and skills with their employees in an effort to help them to advance within the organization.

5.1.4 Job Importance

Job importance refers to an employee's need to feel that his or her skills and abilities are valued and make a difference. Employees that do not feel that they make a difference tend to have difficulty justifying why they should get out of bed in the morning and go to work.

Supervisors should attempt to make every crew member feel like an important part of the team, as if the job wouldn't be possible without their help.

5.1.5 Change

Change refers to an employee's need to have variety in work assignments. Change is what keeps things interesting or challenging. It prevents the boredom that results from doing the same task day after day with no variety.

5.1.6 Personal Growth

Personal growth refers to an employee's need to learn new skills, enhance abilities, and grow as a person. It can be very rewarding to master a new competency on the job, particularly one that you were not trained to do. Similar to change, personal growth prevents the boredom associated with doing the same thing day after day without developing any new skills.

Supervisors should encourage the personal growth of their employees as well as themselves. Learning should be a two-way street on the job site; supervisors should teach their crew members and learn from them as well. In addition, crew members should be encouraged to learn from each other.

5.1.7 Rewards

Rewards refers to the need to be compensated for hard work. Rewards can include a crew member's base salary or go beyond that to include bonuses, give-aways, or other incentives. They can be monetary in nature (salary raises, holiday bonuses, etc.), or they can be non-monetary, such as free merchandise or other prizes (shirts, coffee mugs, jackets, etc.).

5.2.0 Motivating Employees

To increase motivation in the workplace, supervisors must individualize how they motivate different crew members. It is important that supervisors get to know their crew members and determine what motivates them as individuals. Once again, as diversity increases in the workforce, this becomes even more challenging; therefore, effective communication skills are essential.

Here is a list of some tips for motivating employees:

- *Keep jobs challenging and interesting.* Boredom is a guaranteed de-motivator.

- *Communicate your expectations.* People need clear goals in order to feel a sense of accomplishment when they are achieved.

- *Involve your employees.* Feeling that their opinions are valued leads to pride in ownership and active participation.

- *Provide sufficient training.* Give employees the skills and abilities they need to be motivated to perform.

- *Mentor your employees.* Coaching and supporting your employees boosts their self-esteem, their self-confidence, and ultimately their motivation.

- *Lead by example.* Become the kind of leader that employees admire and respect, and they will be motivated to work for you.

- *Treat employees well.* Be considerate, kind, caring, and respectful; treat employees the way that you want to be treated.

- *Avoid using scare tactics.* Threatening employees with negative consequences can backfire, resulting in employee turnover instead of motivation.

- *Reward your crew for doing their best by giving them easier tasks from time to time.* It is tempting to always give your best employees the hardest or dirtiest jobs because you know they will do the jobs correctly.

- *Reward employees for a job well done.*

- *Recognize a job well done and praise it.*

PARTICIPANT EXERCISE B

You are the supervisor of a masonry crew. Sam Williams, an African American, is the person whom the company holds responsible for ensuring that equipment is operable and distributed to the jobs in a timely manner.

Occasionally, disagreements with Sam have resulted in tools and equipment arriving late. Sam, who has been with the company 15 years, resents having been placed in the job and feels that he outranks all the crew supervisors.

Sam figured it was about time he talked with someone about the abuse certain tools and other items of equipment were receiving on some of the jobs. Saws were coming back with guards broken and blades chewed up, bits were being sheared in half, motor housings were bent or cracked, and a large number of tools were being returned covered with mud. Sam was out on your job when he observed a mason carrying a portable saw by the cord. As he watched, he saw the mason bump the swinging saw into a steel column. When the man arrived at his work-station, he dropped the saw into the mud.

You are the worker's supervisor. Sam approached as you were coming out of the work trailer. He described the incident. He insisted, as crew leader, you are responsible for both the work of the crew and how its members use company property. Sam concluded, "You'd better take care of this issue as soon as possible! The company is sick and tired of having your people mess up all the tools!"

You are aware that some members of your crew have been mistreating the company equipment.

1. How would you respond to Sam's accusations?

_____

_____

_____

_____

_____

_____

_____

3. What action would you take regarding the misuse of the tools?

_____

_____

_____

_____

_____

_____

_____

2. Knowing that Sam is an African American, would you respond any differently? Why or why not?

_____

_____

_____

_____

_____

_____

_____

4. How can you motivate the crew to take better care of their tools? Explain.

_____

_____

_____

_____

_____

_____

_____

SECTION 6

6.0.0 TEAM BUILDING

Organizations are making the shift from the traditional boss-worker mentality to one that promotes teamwork. The manager becomes the team leader, and the workers become team members. They all work together to achieve the common goals of the team.

The Society for Human Resources Management (SHRM) reports that there are a number of benefits associated with teamwork. They include the ability to complete complex projects more quickly and effectively, higher employee satisfaction, and a reduction in turnover.

6.1.0 Successful Teams

Successful teams are comprised of individuals who are willing to share their time and talents in an effort to reach a common goal—the goal of the team. Members of successful teams possess an "Us" attitude; they consider what's best for the team and put their egos aside.

Some characteristics of successful teams include the following:

- Everyone participates, and every team member counts
- There is a sense of mutual trust and interdependence
- Team members are empowered
- They communicate
- They are creative and willing to take risks
- The team leader has strong people skills and is committed to the team

6.2.0 Building Successful Teams

To be successful in the team leadership role, the crew leader should contribute to a positive attitude within the team.

There are several ways in which the team leader can accomplish this. First, he or she can work with the team members to create a vision or purpose of what the team is to achieve. It is important that every team member is committed to the purpose of the team, and the team leader is instrumental in making this happen.

Team leaders within the construction industry are typically assigned a crew. However, it is beneficial for the team leader to be involved in selecting the team members. Selection should be based on a willingness of people to work on the team and the resources that they are able to bring to the team.

When beginning a new team, team leaders should do the following:

- *Explain the purpose of the team.* Team members need to know what they will be doing, how long they will be doing it (if they are temporary or permanent), and why they are needed.

- *Help the team establish goals or targets.* Teams need a purpose, and they need to know what it is they are responsible for accomplishing.

- *Define team member roles and expectations.* Team members need to know how they fit into the team and what is expected of them as members of the team.

- *Plan to transfer responsibility.* Teams should be responsible for the tasks to be accomplished. With this responsibility comes a level of autonomy and empowerment to make group decisions.

SECTION 7

7.0.0 GETTING THE JOB DONE

A supervisor is unable to complete all of the activities within a job independently; rather, other people are relied on to get everything done. In addition, supervisors must implement policies and procedures to make sure that the work is done correctly.

7.1.0 Delegating Responsibilities

Construction jobs have crews of people with various experiences and skill levels available to assist in the work. The supervisor's job is to draw from this expertise to get the job done well and in a timely manner.

Once the various activities that make up the job have been determined, the supervisor must identify the person or persons who will be responsible for completing each activity.

This requires that the supervisor be aware of the skills and abilities of the people on the crew. Then, the supervisor must put this knowledge to work in matching the crew's skills and abilities to specific tasks that must be accomplished to complete the job.

Upon matching crew members to specific activities, the supervisor must then delegate the assignments to the responsible person(s). The delegation of responsibilities is generally communicated verbally by the supervisor talking directly to the person that has been assigned the activity. However, there may be times when work is assigned indirectly through written instructions or verbally through someone other than the supervisor.

Some tips for delegating work are:

- Delegate work to a crew member that can do the job properly. If it becomes evident that he or she does not perform to the standard desired, either teach the crew member to do the work correctly or turn it over to someone else who can.

- When delegating, make sure that the crew member understands what to do and the level of responsibility. Make clear the results desired, and specify the boundaries and deadlines for accomplishing the results as well as the resources available.

- Identify the standards and methods of measurement for progress and accomplishment, along with the consequences of achieving or not achieving the desired results. Discuss the task with the crew member, and check for understanding by asking questions. Allow the crew member to contribute feedback or make suggestions about how the task should be performed in a safe and quality manner.

- After delegating the task, give the crew member the time and freedom to get started without feeling the pressure of too much supervision. When making the work assignment, be sure to tell the crew member how much time there is to complete it, and confirm that this time is consistent with the job schedule.

- After a crew member completes a task, examine and evaluate the result. Then, give the crew member some feedback as to how well it has been done. Get the crew member's comments. The information obtained at this time is valuable and will enable the supervisor to know what kind of work to assign that crew member in the future. It will also give the supervisor a means of measuring his or her own effectiveness in delegating work.

7.2.0 Implementing Policies and Procedures

Every company establishes policies and procedures that employees are expected to follow and the supervisors are expected to implement. Company policies and procedures are essentially guides for how the organization does business. They can also reflect organizational philosophies such as putting safety first or making the customer the number-one priority. Examples of policies and procedures include safety guidelines, credit standards, and billing processes.

Here are some tips for implementing policies and procedures.

- Learn the purpose of each policy. That way, you can follow it and apply it properly and fairly.

- If you're not sure how to apply a company policy or procedure, find out.

- Apply company policies and procedures. Remember that they combine what's best for the customer and the company. In addition, they provide direction on how to handle specific situations and answer questions.

- Use common sense when applying company policies or procedures. Some situations may warrant bending a company policy; others must be implemented in order to protect the company (for example, not permitting visitors without hard hats in dangerous construction areas).

Sometimes, supervisors issue orders to their crew members. Basically, an order initiates, changes, or stops an activity. Orders may be general or specific, written or oral, and formal or informal. The decision of how an order will be issued is up to the supervisor, but it is governed by the policies and procedures established by the company.

Some tips on issuing orders are:

- Make orders as specific as possible.

- Avoid giving general or vague orders unless it is impossible to foresee all of the circumstances that could occur in carrying out the order.

- It is not necessary to write orders for simple tasks unless the company requires that all orders be written.

- Write orders for more complex tasks that will take considerable time to complete or orders that are permanent.

- Consider the order, the audience to whom it applies, and the situation under which it will be implemented to determine the appropriate level of formality for the order.

SECTION 8

8.0.0　PROBLEM SOLVING AND DECISION MAKING

Like it or not, problem solving and decision making are a large part of every supervisor's daily work. There will always be problems to be resolved and decisions to be made, especially in fast-paced, deadline-oriented industries such as construction.

8.1.0　Problem Solving vs. Decision Making

Sometimes, the difference between problem solving and decision making is not clear. *Decision making* refers to the process of choosing an alternative course of action in a manner appropriate for the situation. *Problem solving* involves determining the difference between the way things are and the way things should be, and finding out how to bring the two together. The two activities are interrelated because in order to make a decision, you may also have to use problem-solving techniques.

8.2.0　Types of Decisions

Some decisions are routine or simple. These types of decisions can be made based on past experiences. An example would be deciding how to get to and from work. If you've worked at the same place for a long time, you are already aware of the options for traveling to and from work (take the bus, drive a car, carpool with a co-worker, take a taxi, etc.). Based on past experiences with the options identified, you can make a decision of how best to get to and from work.

On the other hand, some decisions are non-routine or more difficult. These types of decisions require more careful thinking about how to carry out an activity by using a formal problem-solving technique. An example is planning a trip to a new vacation spot. If you are not sure how to get there, where to stay, what to see, etc., one option is to research the area to determine the possible routes, hotel accommodations, and attractions. Then, you will have to make a decision about which route to take, what hotel to choose, and what sites to visit, without the benefit of direct past experience with the options.

8.3.0　Formal Problem-Solving Techniques

To make non-routine decisions, the following procedure can be used as a part of formal problem solving.

Step 1　Recognize that a problem exists.

A *problem* is the difference between the way things are and the way that you want them to be. Solving a problem refers to eliminating the differences between what's existing and what's desired, or making things turn out the way you want them.

Consider the following example:

It is 3:30 p.m. Friday afternoon, and you are walking back to the job site trailer after making the afternoon rounds on your job. As you pass the carpenter's workshop, you can't help but notice a lot of good looking, useable lumber in the scrap pile. This bothers you, so you proceed to ask the carpenter in charge why so much good lumber is in the scrap pile. Mary, the head carpenter, states she was not aware that there was so much useable lumber being scrapped.

You both decide to take a closer look at the scrap pile to see just how much is salvageable. You both conclude that a lot of usable lumber is being thrown away. Upon further questioning of the carpenter, you find that she has been sick the last two days, and Mary left her helper, Tom, to do the work.

Investigating even further, you find that Tom just graduated from high school and is enrolled in the first semester of the first-year carpentry program. According to his instructor, Tom appears willing to learn and performs satisfactorily in the classroom. Only the basics of carpentry have been taught to date (math, trade hand tools, and print reading).

Rereading the scenario, the first thing to do to solve the problem is to identify the signals (statements) that indicate that a problem exists. In this case, the main signal is that "a lot of good looking, useable lumber is in the scrap pile."

There may be other areas related to the one noted above. If so, what are they? As the reader can see, the signal is the thing that will motivate the supervisor to find a solution to the problem.

Step 2 Determine what you know or assume you know about the problem, and separate facts from non-facts.

This second step involves answering questions about the problems. Specifically, the *what* (what is the problem? what caused the problem?), *when* (when did it occur?), *where* (where did it happen?), and *who* (who caused the problem? who is involved?) should be determined. Once these questions have been considered, the information must be categorized as factual or non-factual.

Referring to the example above, some responses to an analysis of the problem may be:

What?
- There is too much good, usable lumber in the scrap pile.
- There is too much waste of materials on this job site.
- The job has gone over budget because of this waste.

When?
- The problem was discovered on Friday afternoon.

Where?
- The problem is the scrap pile near the carpenter's workshop.

Who?
- Mary, the carpenter in charge, was off sick for two days, and she left a helper in charge.

- The carpenter acted incorrectly by leaving the helper in charge.
- The helper, Tom, doesn't know how to do his job.

Based on the analysis above, some of the statements are facts while the others are assumptions, judgements, expressions of frustration, and/or blame. To solve the real problem, the supervisor must eliminate the non-factual items and deal only with the facts.

Facts:
- There is too much good, usable lumber in the scrap pile.
- The problem was discovered on Friday afternoon.
- The problem is in the scrap pile near the carpenter's workshop.
- The carpenter in charge was off sick for two days, and she left a helper in charge.

Non-facts:
- There is too much waste of materials on this job site.
- The job has gone over budget because of this waste.
- The carpenter acted incorrectly by leaving the helper in charge.
- The helper doesn't know how to do his job.

Step 3 State the problem.

Using only facts, formulate a problem statement(s). One should note that a fact is a statement of the problem when it meets the following criteria:

- A fact notes the difference between "what is" and "what should be."
- A fact indicates that the present situation has potential for change.

Using the factual statements from the example case, the problem statements are as follows:

- There is much good lumber in the scrap pile that could be used in the construction of the job.
- The carpenter's helper may not know how to cut lumber to minimize waste.

Step 4 Develop objectives that will eliminate the problem.

Describing the desired end result (an objective) is a turning point in the problem-solving sequence. At this point, decision making comes into play. It basically involves choosing to do one of two things: to act on only one idea that comes into mind or to act on what appears to be the most effective of several alternatives.

Written or stated objectives are the basis of the decision-making process. The more precise the objective, the easier it will be to choose the most effective and efficient plan of action.

Continuing with the example above, some of the objectives may be the following:

- By the end of business on Monday, I will have found out whether or not Tom knows how to cut lumber in a way that will reduce waste.

- By the end of business on Tuesday, I will have had a discussion with Mary to ensure that she realizes that training helpers is a part of her job.

- By the end of business on Wednesday, Mary will have prepared an on-the-job training program for Tom.

- By the end of business on Friday, Tom will have begun an on-the-job training program on how to select and cut lumber efficiently.

Step 5 Develop alternate solutions, and select one that will solve the problem.

At this point, the supervisor should list all possible actions that can be taken to accomplish the objectives (solve the problem). When developing the list, judgements should not be made about each alternative. Instead, each possible action should be listed, and then the supervisor should weigh the alternatives to make a decision.

The previous example might have the following list of alternatives:

- Talk to Tom to determine whether or not he is aware of how to cut lumber efficiently; if not, provide training to enable him to do so.

- Provide an on-the-job training program.

- Discuss the problem with Mary and determine the best solution together.

- Inform Mary that it's her responsibility as a lead person to provide training where and whenever needed. Therefore, instruct her to develop and implement a training program.

- Follow up after the program to see if, in fact, Tom can now perform satisfactorily.

Step 6 Develop a plan of action.

Once the solution to the problem has been determined, the next step is to decide who is responsible for carrying out the plan and the deadline for accomplishing it, if not already stated in the objective(s).

For our example, one definite plan of action would be:

1. Meet with Mary to discuss the problem. Get a commitment from her that by Monday she will have discussed the problem with her helper and will report back with the outcome.

2. Upon meeting with Mary the second time, obtain a commitment from her to develop a one-week on-the-job training program for Tom. Inform Mary that it will be her duty to implement the training program and follow up on it. The plan is due by Tuesday.

3. Meet with Mary to review and make any needed revisions to her plan.

4. Follow up next week to ensure that Tom has successfully completed the training program.

Once all six steps of the formal problem-solving technique are completed, the supervisor must follow up to ensure that the action plan was completed as intended. If not, corrective measures should be taken to see that the plan is carried out. In addition, the supervisor should take steps to prevent the same problem from occurring again in the future.

PARTICIPANT EXERCISE C

1. Refering back to the problem example, explain how *you* would handle this situation.

   _____

   _____

   _____

   _____

   _____

   _____

   _____

   _____

   _____

   _____

   _____

   _____

   _____

   _____

   _____

2. Give an example of a decision that you make which does not require a formal problem-solving technique.

   _____

   _____

   _____

   _____

   _____

   _____

   _____

   _____

   _____

   _____

   _____

3. Using the six-step formal problem-solving technique, solve a problem that you are now having or have recently experienced on your job. Write out all the steps.

   _____

   _____

   _____

   _____

   _____

   _____

   _____

   _____

   _____

   _____

   _____

   _____

   _____

   _____

   _____

   _____

   _____

   _____

   _____

   _____

   _____

   _____

   _____

   _____

   _____

   _____

8.4.0 Special Leadership Problems

Because they are responsible for leading others, it is inevitable that supervisors will encounter problems on occasion and be forced to make decisions about how to respond. Some problems will be relatively simple to resolve, like covering for a sick crew member who has taken a day off from work. Other problems will be complex and much more difficult to handle.

There are some complex problems that are relatively common. A few of the major employee problems include:

- Poor attitude towards the workplace
- An inability to work with others
- Absenteeism and turnover

8.4.1 Poor Attitude Towards the Workplace

Sometimes, employees have poor attitudes towards the workplace because of bad relationships with their fellow employees, negative outlooks on their supervision, or a dislike of the job in general. Whatever the case, it is important that the supervisor determine the cause of the poor attitude.

The best way to determine the cause of a poor attitude is to talk with that employee one-on-one, listening to what the employee has to say and asking questions to uncover information. Once this conversation has occurred and the facts have been assessed, the supervisor can determine how to correct the situation and turn the negative attitude into a positive one.

If the supervisor discovers that the problem stems from factors in the workplace or the surrounding environment, the supervisor has several choices. First, the supervisor can move the worker from the situation to a more acceptable work environment. Next, the supervisor can change that part of the work environment found to be causing the poor attitude. Finally, the supervisor can take steps to change the employee's attitude so that the work environment is no longer a negative factor.

8.4.2 Inability to Work with Others

Sometimes supervisors will encounter situations where an employee has a difficult time working with others on the crew. This could be a result of personality differences, an inability

to communicate, or some other cause. Whatever the reason, the supervisor must address the issue and get the crew working as a team.

The best way to determine the reason for why individuals don't get along or work well together is to talk to the parties involved. The supervisor should speak openly with the employee as well as the other individual(s) to find out why.

Once the reason for the conflict is found, the supervisor can determine how to respond. There may be a way to resolve the problem and get the workers communicating and working as a team again. On the other hand, there may be nothing that can be done that will lead to a harmonious situation. In this case, the supervisor would either have to transfer the employee to another crew or have the problem crew member terminated. This latter option should be used as a last measure and should be discussed with one's superiors or Human Resources Department.

8.4.3 Absenteeism and Turnover

Absenteeism and turnover are big problems on construction jobs. Without workers available to do the work, jobs are delayed, and money is lost.

Absenteeism refers to workers missing their scheduled work time on a job. Absenteeism has many causes, some of which are inevitable. For instance, people get sick, they have to take time off for family emergencies, and they have to attend family events such as funerals. However, there are some causes of absenteeism that could be prevented by the supervisor.

The most effective way to control absenteeism is to make the company's policy clear to all employees. Companies that do this find that chronic absenteeism diminishes as a problem. New employees should have the policy explained to them. This explanation should include the number of absences allowed and the reasons for which sick or personal days can be taken. In addition, all workers should know how to inform their supervisors when they miss work and understand the consequences of exceeding the number of sick or personal days allowed.

Once the policy on absenteeism is explained to employees, supervisors must be sure to implement it consistently and fairly. If the policy is administered equally, employees will likely follow it. However, if the policy is not administered equally and some employees are given exceptions, then it will not be effective. Consequently, the rate of absenteeism will increase.

Despite having a policy on absenteeism, there will always be employees who are chronically late or miss work. In cases where an employee abuses the absenteeism policy, the supervisor should discuss the situation directly with the employee. The supervisor should confirm that the employee understands the company's policy and insist that the employee comply with it. If the employee's behavior continues, disciplinary action may be in order.

Turnover refers to the loss of an employee that is initiated by that employee. In other words, the employee quits and leaves the company to work elsewhere or is fired for cause.

Like absenteeism, there are some causes of turnover that cannot be prevented and others that can. For instance, it is unlikely that a supervisor could keep an employee who finds a job elsewhere earning twice as much money. However, supervisors can prevent some employee turnover situations. They can work to ensure safe working conditions for their crew, treat their workers fairly and consistently, and help promote good working conditions. The key to doing so is communication. Supervisors need to know the problems if they are going to be able to successfully resolve them.

Some of the major causes of turnover include:

- *Uncompetitive wages and benefits* — Workers may leave one construction company to go to another that pays higher wages and/or offers better benefits. They may also leave to go to another industry that pays more.
- *Lack of job security* — They leave to find more permanent employment.
- *Unsafe project sites* — They leave for safer projects.
- *Unfair/inconsistent treatment by their immediate supervisor*
- *Poor working conditions*

Essentially, the same actions described above for absenteeism are also effective for reducing turnover. Past studies have shown that maintaining harmonious relationships on the job site will go a long way in reducing both turnover and absenteeism. This will take effective leadership on the part of the supervisor.

PARTICIPANT EXERCISE D

Case I:

On the way over to the job trailer, you look up and see a piece of falling scrap heading for one of the laborers. Before you can say anything, the scrap material hits the ground about five feet in front of the worker. You notice the scrap is a piece of conduit. You quickly pick it up, assuring the worker you will take care of this matter.

Looking up, you see your crew on the third floor in the area from which the material fell. You decide to have a talk with them. Once on the deck, you ask the crew if any of them dropped the scrap. The men look over at Bob, one of the electricians in your crew. Bob replies, "I guess it was mine… it slipped out of my hand."

It is a known fact that the Office of Occupational Safety and Health Administration (OSHA) regulations state that an enclosed chute of wood shall be used for material waste transportation from heights of 20 feet or more. It is also known that Bob and the laborer who was almost hit have been seen arguing lately.

1. Assuming Bob's action was deliberate, what action would you take?

_____
_____
_____
_____
_____

2. Assuming the conduit accidentally slipped from Bob's hand, how can you motivate him to be more careful?

_____

_____

_____

3. What follow-up actions, if any, should be taken relative to the laborer who was almost hit?

_____

_____

_____

4. Should you discuss the apparent OSHA violation with the crew? Why or why not?

_____

_____

_____

5. What acts of leadership would be effective in this case? To what leadership traits are they related?

_____

_____

_____

Case II:

Mike James had just been appointed supervisor of a tile-setting crew. Before his promotion into the management ranks, he had been a tile setter for five years. His work had been consistently of superior quality.

Except for a little good-natured kidding, Mike's co-workers had wished him well in his new job. During the first two weeks, most of them had been cooperative while Mike was adjusting to his supervisory role.

At the end of the second week, a disturbing incident took place. Having just completed some of his duties, Mike stopped by the job site wash station. There he saw Steve and Ron, two of his old friends who were also in his crew, washing.

"Hey Ron, Steve, you should not be cleaning up this soon. It's at least another thirty minutes until quitting time," said Mike. "Get back to your work station, and I'll forget I saw you here."

"Come off it, Mike," said Steve. "You used to slip up here early on Fridays. Just because you have a little rank now, don't think you can get tough with us." To this Mike replied, "Things are different now. Both of you get back to work, or I'll make trouble." Steve and Ron said nothing more, and they both returned to their work stations.

From that time on, Mike began to have trouble as a supervisor. Steve and Ron gave him the silent treatment. Mike's crew seemed to forget how to do the most basic activities.

The amount of rework for the crew seemed to be increasing. By the end of the month, Mike's crew was behind schedule.

1. How do you think Mike should have handled the confrontation with Ron and Steve?

_____

_____

_____

2. What do you suggest Mike could do about the silent treatment he got from Steve and Ron?

_____

_____

_____

3. If you were Mike, what would you do to get your crew back on schedule?

_____

_____

_____

4. What acts of leadership could be used to get the crew's willing cooperation?

_____

_____

_____

5. To which leadership traits do they correspond?

_____

_____

_____

REVIEW QUESTIONS ◆ CHAPTER TWO, SECTIONS 1.0.0–8.4.3

1. A supervisor differs from a craftworker in that _____.

    a. a supervisor need not have direct experience in those job duties that a craftworker typically performs

    b. a supervisor can expect to oversee one or more craftworkers in addition to performing some of the typical duties of the craftworker

    c. a supervisor is exclusively in charge of overseeing, since performing technical work is not part of this role

    d. a supervisor's responsibilities do not include being present on the job site

2. Among the many traits effective leaders should have is _____.

    a. the ability to communicate the goals of a project

    b. the drive necessary to carry the workload by themselves in order to achieve a goal

    c. a perfectionist nature, which ensures that they will not make useless mistakes

    d. the ability to make decisions without needing to listen to others' opinions

3. Of the three styles of leadership, the _____ style would be effective in dealing with a craftworker's negative attitude.

    a. hands-off, followed by autocratic

    b. autocratic, followed by hands-off

    c. democratic

    d. autocratic

4. Verbal communication involves _____.

    a. feedback, in which the receiver of a message should "parrot back" the content of the message in order to let the sender know that the message was understood

    b. the message, which can contain jargon and technical terms used within the profession

    c. the sender, who should remember to concentrate more on the content of the message rather that on the manner in which it was expressed

    d. the receiver, who should first focus on removing barriers to effective listening

5. One way to overcome barriers to effective communication is to _____.

    a. avoid taking notes on the content of the message, since this can be distracting

    b. react objectively, rather than subjectively, to the message

    c. anticipate the content of the message and interrupt if necessary in order to show interest

    d. think about how to respond to the message while listening

6. All the following are things that motivate employees except _____.

    a. setting and achieving goals

    b. receiving monetary and non-monetary incentives

    c. having opportunities to take on additional responsibilities

    d. having a steady, predictable work environment

REVIEW QUESTIONS ◆ CHAPTER TWO, SECTIONS 1.0.0–8.4.3

(Continued)

7. All of the following are ways that a supervisor can motivate employees *except* _____.

   a. have high expectations of employees

   b. give recognition and praise only if it is really merited in order to make the praise more meaningful

   c. use a "one-size-fits-all" approach, since employees are members of a team with a common goal

   d. make a note of what motivated a particular employee in the past, in order to provide the same motivating factor in the future

8. A supervisor can effectively delegate responsibilities by _____.

   a. refraining from evaluating the employee's performance once the task is completed, since it is a new task for the employee

   b. doing the job for the employee to make sure the task is done correctly

   c. allowing the employee to give feedback and suggestions about the new task

   d. communicating information to the employee generally in written form

9. The following are suggestions for implementing policies and procedures *except* _____.

   a. "Do not bend policies and procedures to fit a particular situation"

   b. "Become familiar with the motivation behind each policy"

   c. "Learn how to apply policies and procedures"

   d. "Consider what is best for both the customer and the company"

10. Problem solving differs from decision making in that _____.

   a. problem solving involves identifying discrepancies between the way a situation is and the way it should be

   b. decision making involves separating facts from non-facts

   c. decision making involves eliminating differences

   d. problem solving involves determining an alternative course of action for a given situation

Safety

OBJECTIVES

Upon completion of this chapter, you will be able to:

1. Demonstrate an understanding of the importance of safety.

2. Give examples of direct and indirect costs of workplace accidents.

3. Identify safety hazards of the construction industry.

4. Explain the purpose of the Occupational Safety and Health Act (OSHA).

5. Discuss OSHA inspection programs.

6. Identify the key points of a safety program.

7. List the steps to train employees on how to perform new tasks safely.

8. Identify a supervisor's safety responsibilities.

9. Explain the importance of having employees trained in first aid and Cardio-Pulmonary Resuscitation (CPR) on the job site.

10. Describe the signals of substance abuse.

11. List the essential parts of an accident investigation.

12. Describe the ways to maintain employee interest in safety.

SECTION 1

1.0.0 SAFETY OVERVIEW

The construction industry loses millions of dollars every year because of work-site accidents. Work-related injuries, sickness, and deaths have caused untold human misery and suffering. Project delays and budget overruns from construction injuries and fatalities are huge, and work-site accidents erode the overall morale of the crew.

Construction work has dangers that are part of the job. Examples include working on scaffolds, using cranes in the presence of power lines, and operating heavy machinery. Despite these hazards, experts believe that applying preventive safety measures could drastically reduce the number of accidents.

The job of the crew leader is to carry out the company's safety program and make sure that all workers are performing their tasks safely. To be successful, the crew leader must:

- Be aware of the costs of accidents

- Understand all federal, state, and local governmental safety regulations

- Be involved in training workers in safe-work methods

- Conduct safety meetings

- Get involved in safety inspections, accident investigations, and fire protection and prevention

Crew leaders have the most impact for ensuring that all jobs are performed safely by their crew members. Providing employees with a safe working environment by preventing accidents and enforcing safety standards will go a long way towards maintaining the job schedule and enabling a job's completion on time and within budget. This chapter will present the major duties of a supervisor in relation to job-site safety.

1.1.0 Accident Statistics

The National Institute for Occupational Safety and Health (NIOSH) estimates that more than 7 million people, or 5 percent of the workforce, are currently employed in the construction industry. Each day, construction workers face the risk of falls, machinery accidents, electrocutions, and other potentially fatal occupational hazards.

NIOSH statistics show that more than 1,000 construction workers are killed on the job each year, more fatalities than in any other industry. According to the 1999 survey of the National Census of Fatal Occupational Injuries, falls were the leading cause of deaths in the construction industry. Falls accounted for nearly 50 percent of the fatalities in 1999. Nearly half of the fatal falls occurred from roofs, scaffolds, or ladders. Roofers, structural metal workers, and painters experienced the greatest number of fall fatalities.

The number of fatalities in the construction industry is exceedingly high, but the number of injuries that occur in the workplace is even higher. NIOSH reports that approximately 15 percent of all workers' compensation costs are spent on injured construction workers. The causes of injuries on construction sites include falls, coming into contact with electric current, fires, and mishandling of machinery or equipment. According to NIOSH, back injuries are the leading safety problem in workplaces.

SECTION 2

2.0.0 COSTS OF ACCIDENTS

Occupational accidents are expensive. The American Medical Association reported that the costs due to work-related injuries and illnesses were estimated to be $171 billion per year in the 1990s. These costs affect the employee, the company, and the construction industry as a whole.

Organizations encounter both direct and indirect costs associated with workplace accidents. Examples of direct costs include workers' compensation claims and sick pay; indirect costs include increased absenteeism, loss of productivity, loss of job opportunities due to poor safety records, and negative employee morale attributed to workplace injuries. There are many other related costs involved with workplace accidents. A company can be insured against some of them, but not others. To compete and survive, companies must control these as well as all other employment-related costs.

2.1.0 Insured Costs

Insured costs related to injuries or deaths include the following:

- Compensation for lost earnings
- Medical and hospital costs
- Monetary awards for permanent disabilities
- Rehabilitation costs
- Funeral charges
- Pensions for dependents

Insurance premiums or charges related to property damages include:

- Fire
- Loss and damage
- Use and occupancy
- Public liability

2.2.0 Uninsured Costs

Uninsured costs related to injuries or deaths include the following:

- First aid expenses
- Transportation costs
- Costs of investigations
- Costs of processing reports
- Down time on the job site while investigations are in progress

Uninsured costs related to wage losses include:

- Idle time of workers whose work is interrupted
- Time spent cleaning the accident area
- Time spent repairing damaged equipment
- Time lost by workers receiving first aid
- Costs of training injured workers in a new career

Uninsured costs related to production losses include:

- Product spoiled by accident
- Loss of skill and experience
- Lowered production or worker replacement
- Idle machine time

Associated costs may include the following:

- Difference between actual losses and amount recovered
- Costs of rental equipment used to replace damaged equipment
- Costs of surplus workers used to replace injured workers
- Wages or other benefits paid to disabled workers
- Overhead costs while production is stopped
- Impact on schedule
- Loss of bonus or payment of forfeiture for delays

Uninsured costs related to off-the-job activities include:

- Time spent on injured workers' welfare
- Loss of skill and experience of injured workers
- Costs of training replacement workers

Uninsured costs related to intangibles include:

- Lowered employee morale
- Increased labor conflict
- Unfavorable public relations
- Loss of bid opportunities because of poor safety records
- Loss of client goodwill

SECTION 3

3.0.0 SAFETY REGULATIONS

To reduce safety and health risks and the number of injuries and fatalities on the job, the federal government has enacted some laws and regulations, including the Occupational Safety and Health Act (OSHA) of 1970. Its purpose is "to assure so far as possible every working man and woman in the Nation safe and healthful working conditions and to preserve our human resources".

To promote a safe and healthy work environment, OSHA issues standards and rules for working conditions, facilities, equipment, tools, and work processes. It does extensive research into occupational accidents, illnesses, injuries, and deaths in an effort to reduce the number of occurrences and adverse effects. In addition, OSHA regulatory agencies conduct workplace inspections to ensure that companies follow the standards and rules.

3.1.0 Workplace Inspections

To enforce OSHA, the government has granted regulatory agencies the right to enter public and private properties to conduct workplace safety investigations. The agencies also have the right to take legal action if companies are not in compliance with the Act. These regulatory agencies employ OSHA Compliance Safety and Health Officers (CSHOs), who are chosen for their knowledge in the occupational safety and health field. The CSHOs are thoroughly trained in OSHA standards and in recognizing safety and health hazards.

States with their own occupational safety and health programs conduct inspections. To do so, they enlist the services of qualified state CSHOs.

Companies are inspected for a multitude of reasons. They may be randomly selected, or they may be chosen because of employee complaints, due to an imminent danger, or as a result of major accidents or fatalities.

3.2.0 Penalties for Violations

OSHA has established monetary fines for the violation of their regulations. The penalties are as follows:

- Willful Violations: $59,000 - $70,000

- Repeated Violations: Maximum $70,000

- Serious, Other-Than-
  Serious, Other Specific
  Violations: Maximum $7,000

- OSHA Notice Violation: $1,000

- Failure to Post OSHA
  200 Summary: $1,000

- Failure to Properly Maintain
  OSHA 200 and 100 Logs: $1,000

- Failure to Promptly and Properly
  Report Fatality/Catastrophe: $5,000

- Failure to Permit Access to
  Records under 1904 regulations: $1,000

- Failure to Follow Advance
  Notification Requirements
  Under 1903.6 Regulations: $2,000

- Failure to Abate – for each
  calendar day beyond
  abatement date: Maximum $7,000

- Retaliation Against Individual
  for Filing OSHA Complaint: $10,000

In addition, there can be criminal penalties or personal liability for not complying with OSHA. States may also inflict penalties for any violations to their safety and health programs.

SECTION 4

4.0.0 SAFETY RESPONSIBILITIES

Each employer must set up a safety and health program to manage workplace safety and health and to reduce injury, illness, and fatalities. The program must be appropriate to the conditions of the workplace. It should consider the number of workers employed and the hazards to which they are exposed while at work.

To be successful, the safety and health program must have management leadership and employee participation. In addition, training and informational meetings play an important part in effective programs. Being consistent with safety policies is the key.

4.1.0 Safety Program

The supervisor plays a key role in the successful implementation of the safety program. The supervisor's attitudes towards the program set the standard for how crew members view safety. Therefore, the supervisor should follow all program guidelines and require crew members to do the same.

Safety programs should consist of the following:

- Safety Policies and Procedures
- Hazard Identification and Assessment
- Safety Information and Training
- Safety Record System
- Accident Investigation Procedures
- Appropriate discipline for not following safety procedures

4.2.0 Safety Policies and Procedures

Employers are responsible for following OSHA and state safety standards. Usually, they incorporate OSHA and state regulations into a safety policies and procedures manual. Such a manual is presented to employees when they are hired.

Basic safety requirements should be presented to new employees during their orientation to the company. If the company has a safety manual, the new employee should be required to read it and sign a statement indicating that it is understood. If the employee cannot read, the employer should have someone read it to the employee and answer any questions that arise. The employee should then sign a form stating that he or she understands the information.

It is not enough to tell employees about safety policies and procedures on the day they are hired and never mention them again. Rather, supervisors should constantly emphasize and reinforce the importance of following all safety policies and procedures. In addition, employees should play an active role in determining job safety hazards and find ways that the hazards can be prevented and controlled.

4.3.0 Hazard Identification and Assessment

Safety policies and procedures should be specific to the company. They should clearly present the hazards of the job. Therefore, supervisors should also identify and assess hazards to which employees are exposed. They must also assess compliance with OSHA and state standards.

To identify and assess hazards, OSHA recommends that employers conduct inspections of the workplace, monitor safety and health information logs, and evaluate new equipment, materials, and processes for potential hazards before they are utilized.

Supervisors and employees play important roles in identifying hazards. It is the supervisor's responsibility to determine what working conditions are unsafe and inform employees of hazards and their locations. In addition, they should encourage their crew members to tell them about hazardous conditions. To accomplish this, supervisors must be present and available on the job site.

The supervisor also needs to help the employee be aware of and avoid the built-in hazards that are part of the construction industry. Examples include working on high buildings and in tunnels that are deep underground, on caissons, in huge excavations with earthen walls that have to be buttressed by heavy steel I-beams, and other naturally dangerous projects.

In addition, the supervisor can take safety measures, such as installing protective railings to prevent workers from falling from buildings, as well as scaffolds, platforms, and shoring.

4.4.0 Safety Information and Training

The employer should provide periodic information and training to new and long-term employees. This should be done as often as necessary so that all employees are adequately trained. Special training and informational sessions should be provided when safety and health information changes or workplace conditions create new hazards.

Whenever a supervisor assigns an experienced employee a new task, the supervisor must ensure that the employee is capable of doing the work in a safe manner. The supervisor can accomplish this by providing safety information or training one-on-one or in a group.

The supervisor should do the following:

- Define the task
- Explain how to do the task safely
- Explain what tools and equipment to use and how to use them safely
- Identify the necessary personal protective clothing
- Explain the nature of the hazards in the work and how to recognize them
- Stress the importance of personal safety and the safety of others
- Hold regular safety meetings with the crew's input
- Review Material Safety Data Sheets that may be necessary

4.5.0 Safety Record Systems

OSHA law requires that employers keep records of hazards identified and document the severity of the hazard. The information should include the likelihood of employee exposure to the hazard, the seriousness of the harm associated with the hazard, and the number of exposed employees.

In addition, the employer must document the actions taken or plans for action to control the hazards. While it is best to take actions immediately, it is sometimes necessary to develop a plan, which includes setting priorities and deadlines and tracking progress in controlling hazards.

Employers who are subject to the recordkeeping requirements of the Occupational Safety and Health Act of 1970 must maintain for each establishment a log of all recordable occupational injuries and illnesses.

Logs must be maintained and retained for five years following the end of the calendar year to which they relate. Logs must be available (normally at the establishment) for inspection and copying by representatives of the Department of Labor, the Department of Health and Human Services, or States accorded jurisdiction under the Act. Employees, former employees, and their representatives may also have access to these logs.

A Material Safety Data Sheet (MSDS) is designed to provide both workers and emergency personnel with the proper procedures for handling or working with a substance that may be dangerous. An MSDS will include information such as physical data (melting point, boiling point, flash point, etc.), toxicity, health effects, first aid, reactivity, storage, disposal, protective equipment, and spill/leak procedures. These sheets are of particular use if a spill or other accident occurs.

4.6.0 Accident Investigation Procedures

In the event of an accident, the employer should investigate the cause of the accident and determine how to avoid it in the future. According to OSHA, the employer must investigate each work-related death, serious injury or illness, or incident having the potential to cause death or serious physical harm. The employer should document any findings from the investigation as well as the action plan to prevent future occurrences. This should be done immediately, with photos if possible.

SECTION 5

5.0.0 SUPERVISOR INVOLVEMENT IN SAFETY

To be effective leaders, supervisors must be actively involved in the safety program. Supervisory involvement includes conducting safety meetings and inspections, promoting first aid and fire protection and prevention, preventing substance abuse on the job, and investigating accidents.

5.1.0 Safety Meetings

A safety meeting may be a brief informal gathering of a few employees or a formal meeting with instructional films and talks by guest speakers. The size of the audience and the topics to be addressed determine the format of the meeting. Small, informal safety meetings are typically conducted weekly.

Safety meetings should be planned in advance, and the information should be communicated to all employees affected. In addition, the topics covered within safety meetings should be timely and practical.

5.2.0 Inspections

Supervisors must make regular and frequent inspections to prevent accidents from happening. They must also control the number of accidents that occur. They need to inspect the job sites where their workers perform their tasks. It is recommended that this be done before the start of work each day and during the day at different times.

Supervisors must correct or protect workers from existing or potential hazards in their work areas. Sometimes, supervisors may be required to work in areas controlled by other contractors. Here, the supervisors may have little control over unsafe conditions. In such cases, they should immediately bring the hazards to attention of the contractor at fault, their superior, and the person responsible for the job site.

Supervisors' inspections are only valuable if action is taken to correct what is wrong or hazardous. Consequently, supervisors must be alert for unsafe acts in their work sites. When an employee performs an unsafe action, the supervisors must explain to the employee why the act was unsafe, ask that the employee not do it again, and request cooperation in promoting a safe working environment. The supervisor may decide to document what happened and what the employee was asked to do to correct the situation. It is then very important that supervisors follow-up to make certain that the employee is complying with the safety procedures. Never allow a safety violation to go uncorrected.

5.3.0 First Aid

An accepted definition of *first aid* is "the immediate and temporary care given the victim of an accident or sudden illness until the services of a physician can be obtained." The victim of an accident or sudden illness at a construction site poses more problems than normal since he or she may be working in a remote location. Frequently, the construction site is located far from a rescue squad, fire department, or hospital, presenting a problem in the rescue and transportation of the victim to a hospital. The worker may also have been injured by falling rock or other materials, so special rescue equipment or first-aid techniques are often needed.

The primary purpose of first aid is to provide immediate and temporary medical care to employees involved in accidents, as well as employees experiencing non-work-related health emergencies, such as heart pains or heart attacks. To meet this objective, every supervisor should be aware of the location and contents of first-aid kits available on the job site. Emergency numbers should be posted in the job trailer. The supervisor should be trained in administering and teaching first aid. OSHA has specific requirements for the contents of first aid kits. In addition, OSHA requires that at least one person trained in first aid be present at the job site at all times. In addition, it is recommended that someone on site be certified in CPR.

The employer benefits by having personnel trained in first aid at each job site in the following ways:

1. The immediate and proper treatment of minor injuries may prevent them from developing into more serious conditions. As a result, medical expenses, lost work time, and sick pay will be eliminated or reduced.

2. It may be possible to determine if professional medical attention is needed.

3. Valuable time can be saved when a trained individual prepares the patient for treatment when professional medical care arrives. As a result, lives can be saved.

The American Red Cross and the United States Bureau of Mines provide basic and advanced first aid courses at nominal costs. The American Red Cross trains on both first aid and CPR. The local area offices of these organizations can provide further details regarding the training available.

5.4.0   Fire Protection and Prevention

NIOSH estimates that fires and explosions kill 200 and injure more than 5,000 workers each year. In 1995, it reported that more than 75,000 workplace fires cost businesses more than $2.3 billion. According to the National Safety Council, fires and burns account for 3.3 percent of all occupational fatalities.

The necessity of fire protection and prevention is increasing as new building materials are introduced. Some building materials are highly flammable. They produce great amounts of smoke and gases, which cause difficulties for fire fighters. Other materials melt when ignited and may spread over floors, preventing fire-fighting personnel from entering areas where this occurs.

OSHA has specific standards for fire safety. They require that employers provide proper exits, fire-fighting equipment, and employee training on fire prevention and safety. It is important that supervisors understand and practice these fire-prevention techniques. For more information, consult OSHA guidelines.

5.5.0   Substance Abuse

Unfortunately, drug and alcohol abuse is a reality within our country. *Drug abuse* means inappropriately using drugs, whether they are legal or illegal. Some people use illegal street drugs, such as cocaine or marijuana. Others use legal prescription drugs incorrectly by taking too many pills, using other people's medications, or self-medicating. Others consume alcohol to the point of intoxication.

It is essential that supervisors enforce company policies and procedures about substance abuse. Supervisors must work with management to deal with suspected drug and alcohol abuse and should not handle these situations themselves. These cases are normally handled by the human resources department or designated manager. There are legal consequences of drug and alcohol abuse and the associated safety implications. That way, the business and the employee's safety are protected. If you suspect that an employee is suffering from drug or alcohol abuse, you should immediately contact your supervisor and/or human resources department for assistance.

Construction supervisors have to deal with immediate job safety because of the dangers associated with construction work. It is the supervisor's responsibility to make sure that safety is maintained at all times. This may include removing workers from a work site where they may be endangering themselves or others.

For example, suppose several crew members go out and smoke marijuana or have a few beers during lunch. Then, they return to work to erect scaffolding for a concrete pour in the afternoon. If you can smell marijuana on the crew members' clothing or the alcohol on their breath, you must step in and take action. Otherwise, they might cause an accident that could delay the project or cause serious injury or death.

It is often difficult to detect drug and alcohol abuse. The best way is to search for identifiable effects, such as those mentioned above or sudden changes in behavior that are not typical of the employee. Some examples include:

- Unscheduled absences; failure to report to work on time

- Significant changes in the quality of work

- Unusual activity or lethargy

- Sudden and irrational temper flair-ups

- Significant changes in personal appearance, cleanliness, or health

There are other more specific signs which should arouse suspicions, especially if more than one is exhibited. Among them are:

- Slurring of speech or an inability to communicate effectively

- Shiftiness or sneaky behavior, such as an employee disappearing to wooded areas, storage areas, or other private locations

- Wearing sunglasses indoors or on overcast days to hide dilated or constricted pupils

- Wearing long-sleeved garments, particularly on hot days, to cover marks from needles used to inject drugs

- Attempting to borrow money from co-workers

- The loss of an employee's tools and company equipment

5.6.0 Accident Investigations

There are two times when a supervisor will be involved with an accident investigation. The first time is when an accident, injury, or report of work-connected illness takes place. If present on site, the supervisor should proceed immediately to the place the incident occurred to see that proper first aid is being provided. He or she will also want to make sure that other safety and operational measures are taken to prevent another incident.

After the incident, the supervisor will need to make a formal investigation and submit a report. An investigation looks for the causes of the accident by examining the situation under which it occurred and talking to the people involved. Investigations are perhaps the most useful tool in the prevention of accidents in the future.

For years, a prominent safety engineer was confused as to why sheet metal workers fractured their toes frequently. The supervisor had not made a thorough accident investigation, and the injured workers were embarrassed to admit how the accidents really occurred. It was later discovered they used the metal reinforced cap on their safety shoes as a "third hand" to hold the sheet metal vertically in place when they fastened it. The sheet metal was inclined to slip and fall behind the safety cap onto the toes and cause a fracture. With a proper investigation, further accidents could have been prevented.

There are four major parts to an accident investigation. The supervisor will be concerned with each one. They are:

- Describing the accident

- Determining the cause of the accident

- Determining the parties involved and the part played by each

- Determining how to prevent re-occurrences

SECTION 6

6.0.0 PROMOTING SAFETY

The main way for supervisors to encourage safety is through example. Supervisors should be aware that their behavior sets standards for their crew members. If a supervisor cuts corners on safety, then the crew members may think that it is okay to do so as well.

The key to effectively promote safety is good communication. It is important to plan and coordinate activities and to follow through with safety programs. The most successful safety promotions occur when employees actively participate in planning and carrying out activities.

REVIEW QUESTIONS ◆ CHAPTER THREE, SECTIONS 1.0.0–5.6.0

1. A construction worker performing a job safely involves _____.

   a. conducting safety inspections after an incident occurs to determine what went wrong

   b. undergoing required training after each accident that occurs

   c. being aware of potential risks

   d. not worrying about the costs of accidents, but instead focusing on potential of injury or death

2. With regard to workplace safety, the crew member is held completely responsible for _____.

   a. reporting hazardous conditions

   b. ensuring that everyone is performing their job in a safe manner

   c. conducting safety meetings

   d. enforcing safety standards among the employees

3. In order to ensure workplace safety, the supervisor should _____.

   a. hold formal safety meetings featuring various films and guest speakers

   b. have crew members conduct on-site safety inspections

   c. notify contractors and their supervisor of hazards, in a situation where the job is being performed in an unsafe area controlled by other contractors

   d. hold crew members responsible for making a formal report and investigation following an accident

4. Not following safety measures can cost a company money in the following ways *except* with respect to _____.

   a. funeral costs

   b. compensation for disability

   c. pensions for dependents of those who were injured or killed

   d. employee layoffs

REVIEW QUESTIONS ◆ CHAPTER THREE, SECTIONS 1.0.0–5.6.0

Some activities used by organizations to help motivate employees on safety and help promote safety awareness include:

- Meetings
- Contests
- Recognition and Awards
- Publicity

6.1.0 Meetings

For information on safety meetings, refer to Section 5 of this chapter.

6.2.0 Contests

Contests are a great way to promote safety in the workplace. Examples of safety-related contests include:

- Sponsoring housekeeping contests for the cleanest job site or work area

- Challenging employees to come up with a safety slogan for the company or department

- Having a poster contest that involves employees or their children creating safety-related posters

- Recording the number of accident-free workdays or person-hours

- Giving safety awards (hats, t-shirts, other promotional items or prizes)

One of the positive aspects of contests is their ability to encourage employee participation. It is important, however, to ensure that the contest has a valid purpose. For example, the posters created in a poster contest can be displayed throughout the organization as reminders.

6.3.0   Recognition and Awards

Recognition and awards serve several purposes. Among them are acknowledging and encouraging good performance, building goodwill, reminding employees of safety issues, and publicizing the importance of practicing safety standards. There are countless ways to recognize and award safety. Examples include:

- Supplying food at the job site when a certain goal is achieved

- Providing a reserved parking space to acknowledge someone for a special achievement

- Giving gift items such as t-shirts or gift certificates to reward employees

- Giving plaques to a department or an individual

- Sending a letter of thanks

- Publicly honoring a department or an individual for a job well done

Creativity can be used to determine how to recognize and award good safety on the worksite. The only precautionary measure is that the award should be meaningful and not perceived as a bribe.

6.4.0   Publicity

Publicizing safety is the best way to get the message out to employees. An important aspect of publicity is to keep the message accurate and current. Safety posters that are hung for years on end tend to lose their effectiveness. It is important to keep ideas fresh.

Examples of promotional activities include posters or banners, advertisements or information on bulletin boards, payroll mailing stuffers, and employee newsletters. In addition, merchandise can be purchased that promotes safety, including buttons, hats, t-shirts, and mugs.

PARTICIPANT EXERCISE E

1. What are the two major types of costs caused by accidents?

   _____
   _____

2. List three causes of accidents.

   _____
   _____
   _____

3. What federal government agency regulates job site safety? _____

4. *True or false:* Every supervisor should be trained in first-aid procedures.   T   F

5. List the six steps to informing and training employees how to do a new task.

   _____
   _____
   _____
   _____
   _____
   _____

6. Why are job site inspections important?

   _____
   _____
   _____
   _____
   _____
   _____

7. *True or false:* Investigations may be the most useful tool in preventing future accidents.   T   F

8. List three signs that may suggest substance abuse.

   _____
   _____
   _____

9. List three ways that supervisors can promote safety.

   _____
   _____
   _____

CHAPTER FOUR

Project Control

OBJECTIVES

Upon completion of this chapter, you will be able to:

1. Describe the three phases of a construction project.

2. Define the three types of project delivery systems.

3. Define planning and describe what it involves.

4. Explain why it is important to plan.

5. Describe the two major stages of planning.

6. Explain the importance of documenting one's work.

7. Describe the estimating process.

8. Explain how schedules are developed and used.

9. Identify the two most common schedules.

10. Explain short-interval production scheduling (SIPS).

11. Describe the different costs associated with building a job.

12. Explain the supervisor's role in controlling costs.

13. Illustrate how to control the main resources of a job: materials, tools, equipment, and labor.

14. Define the terms *production* and *productivity* and explain why they are important.

SECTION 1

1.0.0 PROJECT CONTROL OVERVIEW

The contractor, the project manager, and the crew leader each have management responsibilities for their assigned jobs. For example, the contractor's responsibility begins with obtaining the contract, and it does not end until the client takes ownership of the project. Next, the project manager is generally the person with overall responsibility for coordinating the project. Finally, the crew leader is responsible for coordinating the installation of the work of one or more workers, one or more crews of workers within the company and, on occasion, one or more crews of subcontractors.

This chapter provides the necessary information to effectively and efficiently control a project. It examines estimating, planning and scheduling, resources, and cost control.

1.1.0 Construction Projects

Construction projects are made up of three phases. The first is the *Development* phase, the second is the *Planning* phase, and the third is the *Construction* phase.

1.1.1 Development Phase

A new building project begins when an owner has an idea or concept that requires his or her company or community to construct a new facility or add to an existing facility. The development process is the first stage of planning a new building project. This process involves land research and feasibility studies to ensure that the project has merit. Architects or engineers draw the conceptual drawings that define the project graphically. They then provide the owner with sketches of room layouts or elevations and make suggestions about what construction materials should be used.

During the development phase, an estimate of the proposed project will be established, which anticipates all costs or expenses. Once an estimate has been determined, the financing of the project will be discussed with lending institutions. The architects/engineers and/or the owner will start preliminary reviews with government agencies. These reviews include zoning, building restrictions, landscape requirements, and environmental impact studies.

Also during the development phase, the owner must analyze the project's cost and potential to ensure that its costs do not exceed its market value and that the project provides a good return on investment. If the project passes the "value does not exceed cost" criteria, the architect will proceed to the next phase.

1.1.2 Planning Phase

When the architect/engineer begins to develop the preliminary drawings and specifications, other design professionals such as structural, mechanical, and electrical engineers are brought in. They perform the calculations, make a detailed technical analysis, and check details of the project for accuracy.

The design professionals create drawings and specifications. These drawings and specifications will be used to communicate the necessary information to the contractors, subcontractors, suppliers, and workers that contribute to a project.

During the planning phase, the owners hold many meetings to refine estimates, adjust plans to conform to regulations, and secure a construction loan. If the project is a condominium, an office building, or a shopping center, then a marketing program must be developed. In addition, selling of the project must start before actual construction begins.

Next, contract documents to provide the owner with a complete set of drawings, specifications, and bid documents are produced. After the contract documents are complete, the owner will select the method to obtain contractors. The owner may choose to negotiate with several contractors or select one through competitive bidding. In either case, after the owner selects a contractor and awards the contract, construction begins.

1.1.3 Construction Phase

A general contractor enlists the help of mechanical, electrical, elevator, and other specialty contractors, called *subcontractors*, to complete the construction phase. The general contractor will usually perform one or more parts of the construction, and rely on subcontractors for the remainder of the work. The general contractor is responsible for management of all the trades necessary to complete the project.

As construction nears completion, the architect/engineers, owner, and government agencies start their final inspections and acceptance of the project. If the project has been managed by the general contractor, the subcontractors have performed their work, and the architect/engineers have regularly inspected the project to ensure that local codes have been followed, then the inspection procedure can be completed in a timely manner. This results in a satisfied client and a profitable project for all.

On the other hand, if the inspection reveals faulty workmanship, poor design or use of materials, and violation of codes, then the inspection and acceptance will become a lengthy battle and may result in a dissatisfied client and an unprofitable project.

Figure 7 shows a typical project flow.

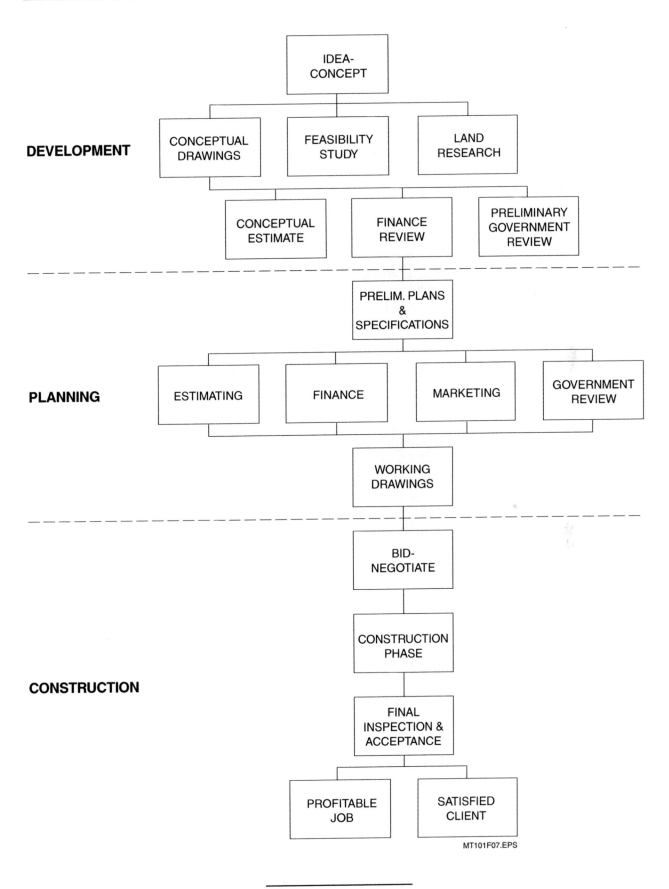

Figure 7 • Project Flow

SECTION 2

2.0.0 PROJECT DELIVERY SYSTEMS

Project delivery systems refer to the process by which projects are delivered, from development through construction. Project delivery systems focus on three main systems: general contracting, design-build, and construction management.

2.1.0 General Contracting

The traditional project delivery system uses a general contractor. In this type of project, the owner determines the design of the project, then solicits proposals from general contractors. After selecting a general contractor, the owner contracts directly with the general contractor, who builds the project as an independent contractor.

2.2.0 Design-Build

The design-build project delivery system is different from the general contracting delivery system. General contracting is not involved in designing the project. In the design-build system both the design and construction of a project are handled by a single contractor.

2.3.0 Construction Management

The construction management project delivery system uses a construction manager to facilitate the design and construction of a project. Construction managers are very involved in project control; their main concerns are controlling time, cost, and quality of the project. There are various forms of construction management. *Figure 8* shows the three main project delivery systems.

SECTION 3

3.0.0 AN OVERVIEW OF PLANNING

3.1.0 What Is Planning?

Planning can be defined as determining the method that will be used to carry out the different tasks to complete a project. It involves deciding what needs to be done and coming up with an organized sequence of events or plan for doing the work.

Planning involves:

- Determining the best method for performing the job
- Identifying the responsibilities of each person on the work crew

	GENERAL CONTRACTING	DESIGN-BUILD	CONSTRUCTION MANAGEMENT
OWNER >	Designs project (or hires architect)	Hires General Contractor	Hires Construction Management Company
GENERAL CONTRACTOR >	Builds project (with owner's design)	Designs project, builds project	Builds, may design (hired by Construction Management Company)
CONSTRUCTION MANAGEMENT COMPANY >			Hires & manages General Contractor and Architect

MT101F08.EPS

Figure 8 • Project Delivery Systems

- Determining the duration of each activity
- Identifying what tools will be needed to complete a job
- Ensuring that the required materials are at the worksite when needed
- Making sure that heavy construction equipment is available when required
- Working with other contractors in such a way as to avoid interruptions and delays

3.2.0 Why Plan?

With a plan, a supervisor can direct work efforts efficiently and can use resources such as personnel, materials, tools, equipment, and work methods to their full potential.

Some reasons for planning include:

- To control the job so that it is built on time and within cost
- To lower job costs
- To prepare for unknowns or unexpected occurrences, such as bad weather, and develop alternate plans
- To promote and maintain favorable employee morale

PARTICIPANT EXERCISE F

1. In your own words, define *planning*, and describe how a job can be done better if it is planned. Give an example.

   _____

   _____

   _____

   _____

   _____

   _____

2. Consider the job that you are working on now to answer the following:

   a. List the material(s) being used.

   _____

   _____

   _____

   _____

   _____

   b. List each member of the crew with whom you work and what each does.

   _____

   _____

   _____

   _____

   _____

   c. List the kinds of equipment being used.

   _____

   _____

   _____

   _____

   _____

   _____

   _____

3. List some suggestions for how your present job could be done better, and describe how you would plan for each of the suggestions.

   _____

   _____

   _____

   _____

   _____

   _____

   _____

   _____

   _____

   _____

SECTION 4

4.0.0 STAGES OF PLANNING

There are various times when planning is done for a construction job. The two most important occur before a project begins, in the pre-construction phase and during the completion of the construction.

4.1.0 Pre-Construction Planning

The pre-construction stage of planning occurs before the start of construction. Except in a fairly small company or for a relatively small job, the supervisor usually does not get directly involved in the pre-construction planning process, but it is important to understand what it involves.

There are two phases of pre-construction planning. The first is when the proposal, bid, or negotiated price for the job is being developed. This is when the estimator, the project manager, and the field supervisors develop a preliminary plan for how the work will be done. This is accomplished by applying experience and knowledge from previous projects. It involves determining what methods, personnel, tools, and equipment will be used and what level of productivity they can achieve.

The second phase occurs after the contractor is awarded the contract. This phase requires a thorough knowledge of any project documents that pertain to the project. During this stage, the actual work methods and resources needed to perform the work are selected. Here, supervisors might get involved, but they must adhere to work methods, production rates, and resources that fit within the estimate prepared before the contract was awarded when planning. If the project requires a method of construction different from what is normal, the supervisor will usually be informed of what method to use.

4.2.0 Construction Planning

During construction, the supervisor is directly involved in planning on a daily basis. This planning consists of selecting methods of completing tasks before beginning them. Effective planning exposes likely difficulties. It enables the supervisor to minimize the unproductive use of personnel and equipment. Effective planning also provides a gauge to measure job progress. Effective supervisors develop what is known as "look-ahead" plan. These are plans that take into consideration actual circumstances as well as projections two to three weeks into the future. Developing this "look-ahead" plan helps to ensure that all resources will be available on the project when needed.

All construction jobs consist of several activities. One of the characteristics of an effective supervisor is the ability to reduce each job to its simpler parts and organize a plan for handling each task.

Others establish time and cost limits for the project, and the supervisor's planning must fit within those constraints. Therefore, it is important to understand the following factors that may affect the work:

- Site and local conditions, such as soil types, accessibility, or available staging areas

- Climate conditions that should be anticipated during the project

- Timing of all phases of work

- Types of materials to be installed and their availability

- Equipment and tools required and their availability

- Personnel requirements and availability

- Relationships with the other contractors and their representatives on the job

On a simple job, these items can be handled almost automatically. However, larger or more complex jobs force the supervisor to give these factors more formal consideration and study.

SECTION 5

5.0.0 THE PLANNING PROCESS

The planning process consists of the following five steps:

1. Establishing a goal
2. Identifying the work activities that must be completed in order to achieve the goal
3. Determining what tasks must be done to accomplish those activities
4. Communicating responsibilities
5. Following up to see that the goal is achieved

5.1.0 Establishing a Goal

A goal is something that needs to be accomplished. There are two types of goals: goals that are set in our personal lives and goals that are set by others in our jobs.

Personal goals are those goals that a person sets for himself or herself. Examples of personal goals include purchasing a new car, moving to a new area, or taking a vacation with friends or family.

Professional or job-related goals are those goals that the organization or project team establishes. Examples of job-related goals include constructing column forms, placing concrete, or setting a water heater for a school building. Although supervisors can provide input on how a job goal will be accomplished, they rarely, if ever, establish the goal.

5.2.0 Identifying the Work to be Done

The second step in planning is to identify the work to be done to achieve the goal. In other words, it is a series of activities that must be done in a certain sequence. The topic of breaking down a job into activities is covered later in this chapter. At this point, the supervisor should know that, for each activity, one or more objectives must be set.

An objective is a statement of what is desired at a specific time. An objective must:

- Mean the same thing to everyone involved
- Be measurable, so that everyone knows when it has been reached
- Be achievable
- Have everyone's full support

Examples of practical objectives include the following:

- By 4:30 p.m. today my crew will have completed laying two-thirds of the 12" interior block wall.
- By closing time Friday, all electrical panels will be installed to standard.

Notice that both examples meet the first three requirements of an objective. In addition, it is assumed that everyone involved in completing the task is committed to achieving the objective. The advantage in writing objectives for each work activity is that it allows the supervisor to evaluate whether or not the plan and schedules are being followed. In addition, objectives serve as subgoals that are usually under the supervisor's control.

Some construction work activities, such as installing 12-inch deep footing forms, are done so often that they require little planning. However, other jobs, such as placing a new type of mechanical equipment, require substantial planning. This type of job requires that the supervisor set specific objectives.

Any time supervisors are faced with a new or complex activity, they should take the time to establish objectives that will serve as guides for accomplishing the task at hand. These guides can be used in the current situation, as well as in similar situations in the future.

5.3.0 Determining Tasks

To plan effectively, the supervisor must be able to break a work activity down into the smaller tasks. Large jobs include a greater number of tasks than small ones, but all jobs can be broken down into manageable components.

When breaking down an activity into tasks, the supervisor should make each task identifiable and definable. A task is identifiable when you know the types and amounts of resources it requires. A task is definable if you can assign a specific time to it. For purposes of efficiency, the supervisor's job breakdown should not be too detailed or complex, unless the job has never been done before or must be performed with strictest efficiency.

A suitable breakdown for the work activity, *install 12-inch × 12-inch vinyl tile in a cafeteria,* might be the following:

Step 1	Lay out
Step 2	Clean floor
Step 3	Spread adhesive
Step 4	Lay tile
Step 5	Clean floor
Step 6	Wax floor

The supervisor could create even more detail by breaking down any one of the tasks, such as *lay tile,* into subtasks. In this case, however, that much detail is unnecessary and wastes the supervisor's time and the project's money. However, breaking tasks down further might be necessary in a case where the job is very complex or the analysis of the job needs to be very detailed.

Every work activity can be divided into three general parts:

- Preparing
- Performing
- Cleaning up

One of the most frequent mistakes made in the planning process is forgetting to prepare and to clean up. The supervisor must be certain that preparation and cleanup are not overlooked with every job breakdown.

After identifying the various activities that make up the job and developing an objective for each activity, the supervisor must determine what resources the job requires. Resources include labor, equipment, materials, and tools. In most jobs, these resources are identified in the job estimate, and the supervisor must only make sure that they are on the site when needed. On other jobs, however, the supervisor may have to determine, in part, what is required as well as arrange for timely delivery.

5.4.0 Communicate Responsibilities

A supervisor is unable to complete all of the activities within a job independently. Other people must be relied on to get everything done. Therefore, construction jobs have a crew of people with various experiences and skill levels to assist in the work. The supervisor's job is to draw from this expertise to get the job done well and in a timely manner.

Once the various activities that make up the job have been determined, the supervisor must identify the person or persons who will be responsible for completing each activity. This requires that the supervisor be aware of the skills and abilities of the people on the crew. Then, the supervisor must put this knowledge to work in matching the crew's skills and abilities to specific tasks that must be accomplished to complete the job.

Upon matching crew members to specific activities, the supervisor must then communicate the assignments to the crew. Communication of responsibilities is generally handled verbally; the supervisor often talks directly to the person to which the activity has been assigned. There may be times when work is assigned indirectly through written instructions or verbally through someone other than the supervisor. Either way, the crew members should know what it is they are responsible for accomplishing on the job.

5.5.0 Follow-Up

Once the activities have been delegated to the appropriate crew members, the supervisor must follow up to make sure that they are completed effectively and efficiently. Follow-up involves being present on the job site to make sure that all of the resources are available to complete the work; ensuring that the crew members are working on their assigned activities; answering any questions; or helping to resolve any problems that occur while the work is being done. In short, follow-up means that the supervisor is aware of what's going on at the job site and is doing whatever is necessary to make sure that the work is completed as scheduled.

Figure 9 reviews the planning steps.

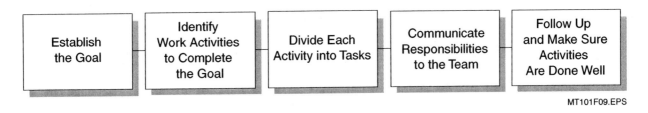

Figure 9 • Steps to Effective Planning

MT101F09.EPS

SECTION 6

6.0.0 PLANNING RESOURCES

Once a job has been broken down into its tasks or activities, the next step is to assign the various resources needed to perform them.

6.1.0 Planning Materials

The materials required for the job are identified during pre-construction planning and are listed on the job estimate. The materials are ordered from suppliers who have previously provided quality materials on schedule and within estimated cost.

The supervisor is usually not involved in the planning and selection of materials, since this is done in the pre-construction phase. However, the supervisor is involved in planning materials for tasks such as job-built formwork and scaffolding. Also, the supervisor occasionally may run out of a specific material, such as fasteners, and need to order more. In such cases, the next higher supervisor should be consulted, since most companies have specific purchasing policies and procedures.

The supervisor does, however, have a major role to play in the receipt, storage, and control of the materials after they reach the job site.

6.2.0 Planning Equipment

Planning the use of construction equipment involves identifying the types of equipment needed, the tasks each type must perform, and the time each piece is needed. Much of this is planned during the pre-construction phase; however, it is up to the supervisor to work with the home office to make certain that the equipment reaches the job site on time.

The supervisor should also coordinate equipment with other contractors on the job. Sharing equipment can save time and money and avoid duplication of effort. In addition, if the equipment breaks down, the supervisor must know who to contact to resolve the problem. The supervisor should also designate some time for routine equipment maintenance to avoid equipment failure.

Finally, when various pieces of equipment will be working in a relatively small area, or one piece will work in conjunction with another (such as a loader and a dump truck, or a bulldozer and a scraper), the supervisor must coordinate the use of the equipment. An alternate plan must be ready in case one piece of equipment breaks down, so that the other equipment does not sit idle. This planning should be done in conjunction with the home office or the supervisor's immediate superior.

6.3.0 Planning Tools

A supervisor is responsible for planning what tools will be used on the job. This includes:

- Determining the tools required

- Informing the workers who will provide the tools (company or worker)

- Instructing the workers how to use the tools safely and effectively

- Determining what controls to establish for tools

6.4.0 Planning Labor

All tasks require some sort of labor because the supervisor cannot complete all of the work alone.

When planning labor, the supervisor must:

- Identify the skills needed to perform the work.

- Determine how many people having those specific skills are needed.

- Decide who will actually be on the crew.

In many companies, the project manager or job superintendent determines the size and makeup of the crew. Then, the supervisor is expected to accomplish the goals and objectives with the crew provided. Even though the supervisor may not be involved in staffing the crew, the supervisor will be responsible for training the crew members to ensure that they have the skills necessary to do the job.

In addition, the supervisor is responsible for keeping the crew adequately staffed at all times so that jobs are not delayed. This involves dealing with absenteeism and turnover, two common problems that affect the construction industry today.

SECTION 7

7.0.0 WAYS TO PLAN

It is recommended that supervisors carry a small note pad to be used for planning and note taking. That way, thoughts about the project can be recorded as they occur, and pertinent details will not be forgotten. The supervisor may also choose to use a planning form, such as the one illustrated in *Figure 10*.

DAILY WORK PLAN

"PLAN YOUR WORK AND WORK YOUR PLAN = EFFICIENCY"

Plan of _____   Date _____

PRIORITY	DESCRIPTION	✓ When Completed ✕ Carried Forward

Figure 10 • Planning Form

As the job is being built, the supervisor should refer back to these plans and notes to see that the tasks are being done in sequence and according to plan. This task is referred to as *analyzing the job*. Experience shows that jobs that are not built according to work plans usually end up costing more and taking more time; therefore, it is important that supervisors refer back to the plans periodically.

The supervisor is involved with many activities on a day-to-day basis. As a result, it is easy to forget important events if they are not documented. To help keep track of events such as job changes, interferences, and visits, the supervisor should keep a job diary.

A job diary is a small book in which the supervisor records activities or events that take place on the job site which may be important at a later date. When recording in a job diary, the supervisor should make sure that the information is accurate, factual, complete, consistent, organized, and up-to-date.

The supervisor should also be sure to follow company policy in determining which events should be recorded. However, if there is a doubt about what to include, it is better to have too much information than too little.

Figure 11 illustrates a sample page from a job diary.

July 8, 2001

Weather: Hot and Humid

Project: Company XYZ Building

- The paving contractor crew arrived late (10am).

- The owner representative inspected the footing foundation at approximately 1pm.

- The concrete slump test did not pass. Two trucks had to be ordered to return to the plant, causing a delay.

- John Smith had an accident on the second floor. I sent him to the doctor for medical treatment. The cause of the accident is being investigated.

Figure 11 • Sample Page from a Job Diary

SECTION 8

8.0.0 ESTIMATING

Before a project is built, an estimate of the cost needs to be prepared. An estimate is the process of calculating the quantities of costs needed to build the job. There are two types of costs to consider, direct and indirect costs.

Direct costs include:

- materials
- labor
- tools
- equipment

Indirect costs refer to overhead items such as:

- office rent
- utilities
- job site telephones
- office supplies signs

The bid price includes the estimated cost of the project as well as the profit. Profit refers to the amount of money that the contractor will make after all of the direct and indirect costs have been paid. If the direct and indirect costs exceed those estimated to perform the job, the difference between the actual and estimated costs must come out of the profit. This reduces what the contractor makes on the job.

Contractors cannot afford to lose on profit regularly; otherwise, they will be forced to go out of business. Therefore, it is very important that supervisors adhere to the drawings, specifications, and job schedule as directed to reduce the chance of the company losing money on the project.

The contractor's main office will develop the actual project estimate. However, the supervisor will have to order materials from time to time on the job. This will require the preparation of a quantity takeoff or material estimate.

8.1.0 The Estimating Process

A complete estimate is developed as follows:

1. Using the drawings and specifications, an estimator records the quantity of the materials needed to construct the job. This is called a *quantity takeoff*. The information is placed on a Quantity Takeoff Sheet (sometimes called a Work Sheet), such as that shown in *Figure 12*.

2. The second step is to calculate the amount of time it takes to perform an amount of work. These are known as *production rates*. Most companies keep records of these rates for the type and size of the jobs that they perform. The estimating department in a company has these figures.

3. Equipment production rates are obtained from company records.

4. The total material quantities are taken from the Quanity Takeoff Sheet and placed on a Summary or Pricing Sheet, an example of which is shown in *Figure 13*. Material prices are obtained from local suppliers, and the total cost of materials is calculated.

5. The labor and production rates, the costs of owning and operating (or renting) the equipment, and the local wage rates are added to obtain the total labor and equipment costs. This information is placed on the Summary Sheet.

6. The total cost of all resources—materials, equipment, tools, and labor—is then totaled on the Summary Sheet. The unit cost—the total cost divided by the total number of units of material to be put into place—can also be calculated.

7. The cost of materials, equipment, tools, and labor for each activity is then placed on a Recapitulation Sheet, as illustrated in *Figure 14*. The Recapitulation Sheet serves as a summary for the entire project.

8. The cost of taxes, bonds, insurance, subcontractors' work, and other indirect costs are added to the direct costs of the materials, equipment, tools, and labor. This total is added to the contractor's expected profit.

WORK SHEET

Takeoff By: _____

Checked By: _____

DATE _____

SHEET _____ of _____

PROJECT _____

ARCHITECT _____

PAGE #

REF.	DESCRIPTION	NO.	DIMENSIONS			EXTENSION	QUANTITY	UNIT	TOTAL		REMARKS
			LENGTH	WIDTH	HEIGHT				QUANTITY	UNIT	

MT101F12.EPS

Figure 12 • Quantity Takeoff Sheet

SUMMARY SHEET

By:

DATE _____

SHEET _____ of _____

PAGE #

TITLE: _____

PROJECT _____

WORK ORDER # _____

DESCRIPTION	QUANTITY		MATERIAL COST		LABOR MAN HOURS FACTORS					LABOR COST		ITEM COST	
	TOTAL	UT	PER UNIT	TOTAL	CRAFT	PR UNIT	TOTAL	RATE	COST PR	PER	TOTAL	TOTAL	PER UNIT
	MATERIAL									LABOR	TOTAL		

Figure 13 • Pricing Sheet

MT101F18.EPS

Listed By: _____ Checked By: _____	RECAPITULATION SHEET PROJECT _____ DATE _____ ARCHITECT_____ S/SF _____					Page _____ of _____
PAGE REF.	ITEM	MATERIAL	LABOR	SUB	EQUIPMENT	TOTAL

MT101F14.EPS

Figure 14 • Recapitulation Sheet

8.1.1 Estimating Material Quantities

The crew leader may be required to estimate quantities of materials.

To estimate the amount of a certain type of material required to perform a job, a set of construction drawings and specifications will be needed. The appropriate section of the technical specifications and page(s) of drawings should be carefully reviewed to determine the types and quantities of materials required. The quantities are then placed on the Work Sheet. For example, the specification section on finished carpentry should be reviewed along with the appropriate pages of drawings before taking off the linear feet of door and window trim.

If an estimate is required because not enough materials were ordered to complete the job, the estimator must also determine how much more work is necessary. Once this is known, the supervisor can then determine specifically how many more materials are needed. The construction drawings will also be used in this process.

For example, assume you are the supervisor of a crew building footing formwork for the construction shown in *Figure 15*.

You have used all of the materials provided for the job, yet you have not completed it. You study the drawings and see that the formwork is comprised of two side forms, each 12" high. The total length of footing for the entire project is 115'-0". You have completed 88'-0" to date; therefore, you have 27'-0" remaining (115'-0" – 88'-0" = 27'-0"). Your job is to prepare an estimate of materials that you will need to complete the job. In this case, only the side forms will be estimated (the miscellaneous materials will not be considered here).

- Footing length to complete: 27'-0"
- Footing height: 1'-0"

Refer to the Work Sheet in *Figure 16* for a final tabulation of the side forms needed to complete the job.

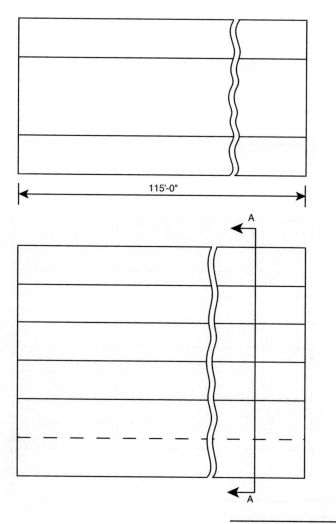

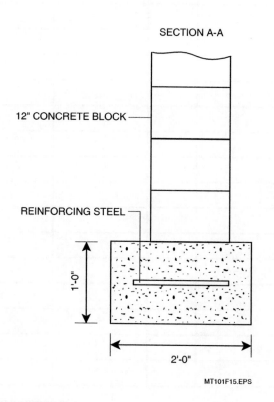

Figure 15 • Footing Formwork

WORK SHEET

PAGE #1

Takeoff By: RWH

Checked By:

DATE __2/1/01__

SHEET __01__ of __01__

PROJECT __Sam's Diner__

ARCHITECT __654b__

| REF. | DESCRIPTION | NO | DIMENSIONS | | | EXTENSION | QUANTITY | UNIT | TOTAL | | REMARKS |
			LENGTH	WIDTH	HEIGHT				QUANTITY	UNIT	
	Footing Side Forms	2	27'0"		1'0"	2x27x1	54	SF	54	SF	

MT101F16.EPS

Figure 16 • Work Sheet With Entries

PARTICIPANT EXERCISE G

1. Using the same footing as described in the example above, calculate the quantity (square feet) of formwork needed to finish 203 linear feet of the footing. Place this information directly on the Work Sheet on the previous page.

2. You are the supervisor of a carpentry crew whose task is to side a warehouse with plywood sheathing. The wall height is 16 feet, and there is a total of 480 linear feet of wall to side. You have done 360 linear feet of wall and have run out of materials. Calculate how many more feet of plywood you will need to complete the job. If you are using 4' × 8' plywood panels, how many will you need to order, assuming no waste? Write your estimate on the Work Sheet on the previous page.

PARTICIPANT EXERCISE G ◆ PROJECT CONTROL

SECTION 9

9.0.0 SCHEDULING

Planning and scheduling are closely related and are both very important to a successful construction job. *Planning* involves determining which activities must be completed and how they should be accomplished. *Scheduling* involves establishing start and finish times or dates for each activity.

When scheduling a job, a construction schedule is generally completed. A construction schedule for a project shows:

- Operations listed in sequential order
- Units of construction
- Duration of activities
- Estimated date to start and complete each activity
- Quantity of materials to be installed

There are two major types of construction schedules used today. The first is the Bar Chart, and the second is the Network Schedule, which is sometimes called the Critical Path Method (CPM) or Precedence Diagram.

9.1.0 The Scheduling Process

The following is a brief summary of the steps required to develop a schedule.

Step 1 Make a list of all of the activities that will be performed to build the job, including individual work activities and special tasks such as inspections or the delivery of materials.

At this point, the supervisor should just be concerned with generating a list, not with determining how the activities will be accomplished, who will perform them, how long they will take, or in what sequence they will be completed.

Step 2 Use the list of activities created in Step 1 to reorganize the work activities into a logical sequence.

When doing this, the supervisor should keep in mind that certain steps cannot happen until others have been completed. For example, footing must be excavated before concrete can be placed.

Step 3 Assign a duration or length of time that it will take to complete each activity and determine the start time for each.

Each activity will then be placed into a final format, known as a schedule. This step is important because it helps the supervisor to ensure that the activities are being completed on schedule.

9.2.0 Bar Charts

Bar charts can be used for both short-term and long-term jobs. However, they are especially helpful for jobs of short duration.

Bar charts provide management with the following:

- A visual concept of the overall time required to complete the job through the use of a logical method rather than a calculated guess
- A means to review each part of the job
- Coordination requirements between crafts
- Alternative methods of performing the work

A bar chart can be used as a control device to see whether or not the job is on schedule. If the job is not on schedule, immediate action can be taken in the office and the field to correct the problem and ensure that the activity is completed on schedule.

A bar chart is illustrated in *Figure 17*.

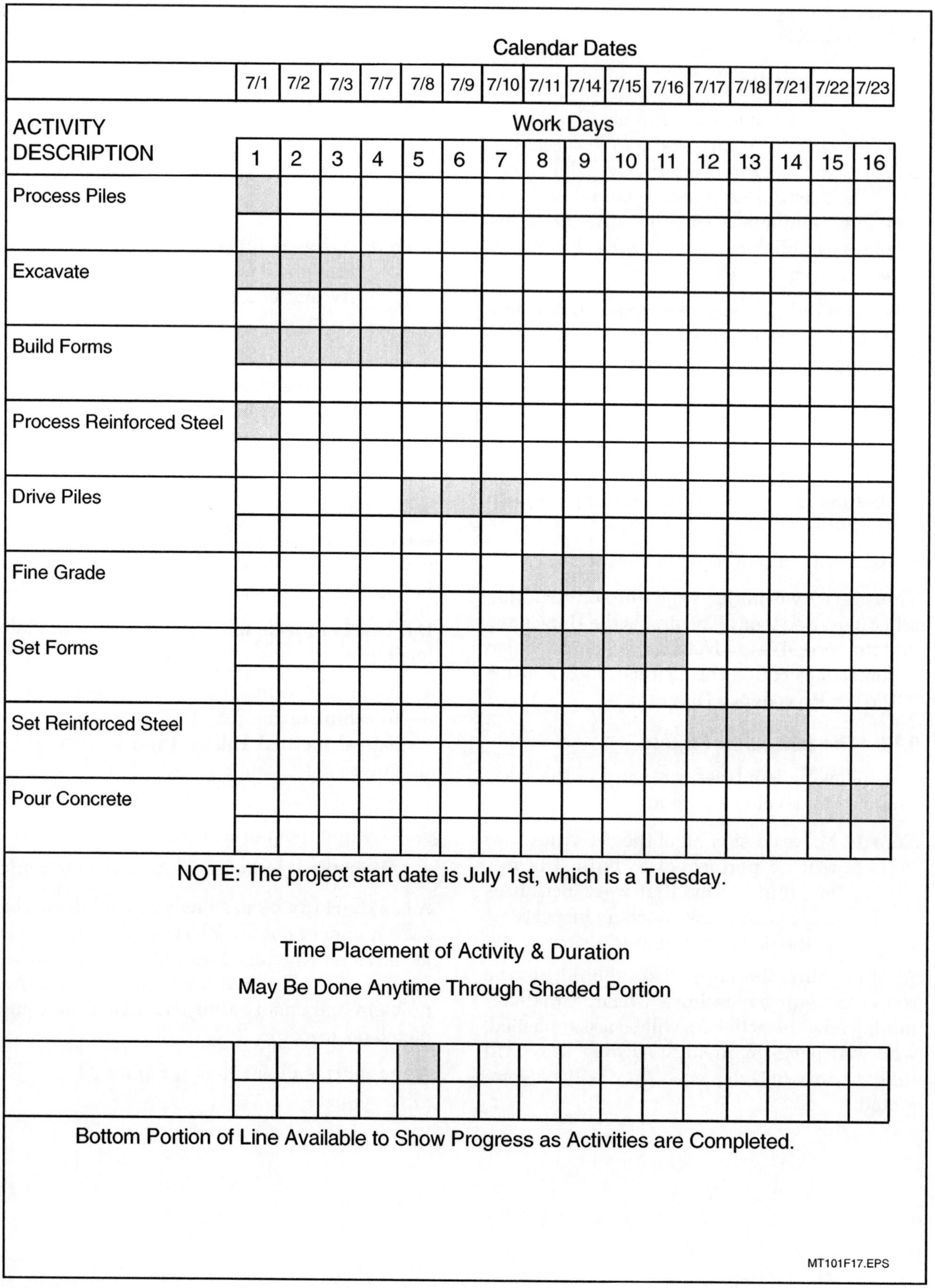

	Calendar Dates															
	7/1	7/2	7/3	7/7	7/8	7/9	7/10	7/11	7/14	7/15	7/16	7/17	7/18	7/21	7/22	7/23
ACTIVITY DESCRIPTION	Work Days															
	1	2	3	4	5	6	7	8	9	10	11	12	13	14	15	16
Process Piles	▓															
Excavate	▓	▓	▓													
Build Forms	▓	▓	▓	▓												
Process Reinforced Steel	▓															
Drive Piles					▓	▓	▓									
Fine Grade								▓	▓							
Set Forms								▓	▓	▓	▓					
Set Reinforced Steel												▓	▓	▓		
Pour Concrete															▓	▓

NOTE: The project start date is July 1st, which is a Tuesday.

Time Placement of Activity & Duration

May Be Done Anytime Through Shaded Portion

Bottom Portion of Line Available to Show Progress as Activities are Completed.

MT101F17.EPS

Figure 17 • Bar Chart

9.3.0 Network Schedule

Network scheduling is a more sophisticated method of scheduling than the bar chart. The network schedule is similar to a road map. Network schedules display many items, such as material or engineering bottlenecks, field problems, completion dates of various job phases, and overall completion times.

Network schedules are generally used for complex jobs that take a long time to complete. Unlike the bar chart, network schedules show dependencies. However, they are not as easy to use as the bar chart.

Refer to *Figure 18* for an example of a network schedule.

9.4.0 Short-Interval Production Scheduling

Since the supervisor needs to maintain the job schedule, he or she needs to be able to plan daily production. Short-Interval Production Scheduling (SIPS) is a method used to do this.

SIPS is defined as a technique for scheduling work of a particular job over a short period of time so that the right amount of work can be accomplished within the estimated costs. The information for SIPS comes from the estimate or estimate breakdown. SIPS helps to translate estimate data and the various job plans into a day-to-day schedule of events.

The purpose of SIPS is to provide short-term schedule information. Its key importance is to compare actual production with estimated production. If actual production begins to slip behind estimated production, SIPS will warn the supervisor that a problem lies ahead and that a schedule over-run is developing.

Furthermore, SIPS can be used to set production goals. It is generally agreed that production can be improved if workers:

- Know the amount of work to be accomplished
- Know the time they have to complete the work
- Have something to say about setting goals

Consider the following example:

Situation: A carpentry crew on a retaining wall project is about to form and pour catch basins and put up wall forms. The crew has put in a number of catch basins, so the supervisor is sure that they can perform the work within the estimate. However, the supervisor is concerned about their production of the wall forms. The crew will work on both the basins and the wall forms at the same time.

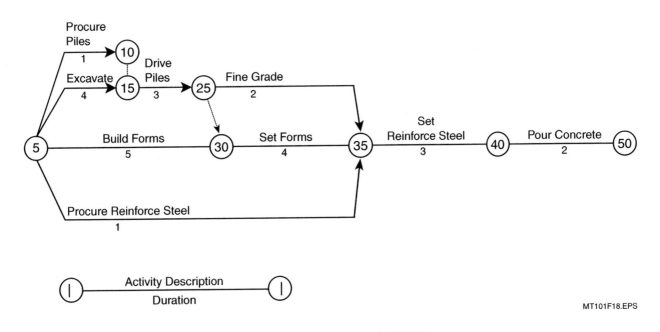

MT101F18.EPS

Figure 18 • Network Schedule

1. The supervisor notices the following in the estimate or estimate breakdown:

    a. Production factor for wall forms = 16 work-hours per 100 square feet

    b. Work to be done by measurement = 800 square feet

    c. Total time = 128 work-hours (800 × 16 ÷ 100)

2. The carpenter crew consists of the following:

    a. One carpenter supervisor

    b. Four carpenters

    c. One laborer

3. The supervisor determines the goal for the job should be set at 128 work-hours (from *Step 1*).

4. If the crew remains the same (six workers), the work should be completed in about 21 crew-hours (128 work hours ÷ 6 workers = 21.33 crew-hours).

5. The supervisor then discusses the production goal (completing 800 square feet in 21 crew-hours) with the crew and encourages them to work together to meet the goal of getting the forms erected within the estimated time.

Therefore, SIPS is needed to translate production into work-hours or crew-hours and to schedule work so that the crew can accomplish it within the estimate. Also, the establishment of a production target or goal provides the motivation to produce more than the estimate requires.

The supervisor's progress will be based on productivity, which is a measure of how well the crew maintains the schedule. The supervisor must be able to read and interpret the job schedule. On some jobs, the beginning and expected end date for each activity along with the expected crew or worker's production rate will be provided on the form. The supervisor can use this information to plan work more effectively, to realistically set goals for the crew, and to measure whether or not they were accomplished within the scheduled time.

It is important, therefore, that all job resources are on the job, in location, and ready to begin the job on time and stay on schedule. Before starting an activity, the supervisor must know or determine what materials, tools, equipment, and labor are needed to complete the job. Next he or she must know or determine when the various resources are needed. Finally, the supervisor must follow up to ensure that they are available on the job site.

The supervisor should not wait until the day of or the day before a job starts to check on the availability of needed resources. Rather, this should be done at least three to four working days before the start of the job. It should be done even earlier for larger jobs. This advance preparation will help prevent any events that could potentially delay the job or cause it to fall behind schedule.

9.5.0 Updating a Schedule

No matter what type of schedule is used, it must be kept up-to-date to be useful to the supervisor. Schedules that are inaccurate offer no value.

The person responsible for scheduling in the office handles the updates. This person uses information gathered from job field reports to update the schedule.

The supervisor is usually not directly involved in updating schedules. However, he or she may be responsible for completing field or progress reports used by the company. It is critical that the supervisor fill out any required forms or reports completely and accurately so that the schedule can be updated with the correct information.

REVIEW QUESTIONS ◆ CHAPTER 4, SECTIONS 9.0.0–9.5.0

1. One of the advantages and disadvantages of the bar chart and the network schedule is that _____.

    a. the bar chart is particularly good for short-term jobs, but is harder to use than the network schedule

    b. the bar chart is good for long-term jobs, but it is harder to use than the network schedule

    c. the network schedule is more sophisticated than the bar chart, but does not indicate dependencies like the bar chart

    d. the network schedule is more sophisticated because it can indicate dependencies, but it is more difficult to use

2. Short-interval scheduling can help crew leaders in the field by _____.

    a. offering a comparison of actual production to estimated production

    b. providing mainly short-term, but also long-term, schedule information

    c. determining the equipment and materials necessary to complete a task

    d. providing the information necessary for coming up with an estimate or an estimate breakdown

3. One way for supervisors to get a job back on schedule or to increase productivity if things fall behind is to _____.

    a. refrain from spending money on hiring more crew members

    b. make sure the necessary tools, materials, and equipment are on the job site a few hours in advance

    c. communicate to crew members the exact amount of time available to complete a task

    d. set goals for crew members instead of directly involving them in the process

4. The following describes how schedules are typically kept up-to-date: _____.

    a. the supervisor typically handles schedule updates

    b. the supervisor typically has someone else handle schedule updates

    c. information from job field reports or progress reports is not typically used for schedule updates

    d. information from bar charts and network schedules is used for schedule updates

REVIEW QUESTIONS ◆ CHAPTER FOUR, SECTIONS 9.0.0–9.5.0

SECTION 10

10.0.0 COST AWARENESS AND CONTROL

Being aware of costs and controlling them is the responsibility of every employee on the job, and it is the supervisor's job to ensure that employees uphold this responsibility. Control refers to the comparison of estimated performance against actual performance and following up with any needed corrective action. Supervisors who use cost-control practices are more valuable to the company than those who do not.

On a typical construction site, hundreds of activities are going on at the same time. The more going on, the harder it is to control the activities involved or constructing the building. It is therefore very important that the construction crew leader be constantly aware of the costs of a project and efficiently and effectively control the various resources used on the job.

When resources are not controlled, the cost of the job increases. For example, a plumbing crew of four people is installing soil pipe and does not have enough wyes. Three crew members wait while one crew member goes to the supply house for a part that costs only a few dollars. It takes the crew member an hour to get the part, so four hours of productive work were lost. In addition, the travel cost for retrieving the supplies must be added.

10.1.0 Categories of Costs

There are two general categories of costs associated with every job, *estimated costs* and *actual costs*. Although different, they are related in determining whether a job was built efficiently.

Estimated costs, as described in Section 8 of this chapter, are calculated before construction begins. These are the costs that are required to build the job. The estimated cost is the contract cost agreed between the contractor and the client or the general contractor and a specialty contractor to do the work.

The estimated project cost is a sum of all individual jobs and estimated costs. Estimates for each job serve as measures of what each should cost when completed. It allows the contractor to measure whether or not the job is being built for the estimated cost. It is similar to using a measuring tape to check a measurement of a piece of work in the field. If it was installed correctly, the measurement should be the same as that indicated on the drawings. If it was not, the work will have to be corrected to achieve the proper measurement.

The same is true of costs; if the job was built for the estimated cost, then the contractor will receive the estimated profit. The company will have been successful and will grow, providing job security for the supervisor. However, if the job is not built for the estimated cost, the contractor will not make the expected profit. If loss continues, the company may have to lay off employees.

Actual costs are the costs for which the contractor is responsible, such as the cost of the delivered materials and the payroll for the employees. Again, it is important that the supervisor remember that a major responsibility is to build the job so that actual costs of doing the work are below or equal to the estimated costs. This should be accomplished without sacrificing the quality required in the contract drawings or specifications.

10.2.0 Field Reporting System

The total estimated cost comes from the job estimates, but the actual cost of doing the work is obtained from an effective field reporting system. A field reporting system is made up of a series of forms, which are completed by different people, including the crew leader. Each company has its own forms and methods for obtaining information. The new supervisor needs to find out what his or her company's methods are. The information and the process of how they are used are described below.

First, the number of hours each person worked on each task must be known. This information comes from daily time cards. Once the accounting department knows how many hours each employee worked on each activity, it can calculate the total cost of the labor by multiplying the number of hours worked by the wage rate for each worker. The cost for the labor to do each task will be calculated as the job progresses, and it will be compared with the estimated cost. This comparison will also be done at the end of the job.

When material is put in place, a designated person in the company will measure the quantities from time to time and send this information to the home office and possibly the supervisor. Knowing this information, along with the actual cost of the material and the amount of hours it took the various workers to install it, a comparison will be made to determine if the actual cost of installing a specific material is below, above, or equal to the estimate. Obviously, if the cost is greater than the estimate, management and the crew leader will have to take action to reduce the cost to get it below or equal to the estimate.

A similar process is used to determine if the costs to operate equipment or the production rate are above, below, or equal to the estimated cost and production. Forms are available for obtaining this information.

If this comparison process is to be of use, the information obtained from field personnel must be correct. It is important that the supervisor be accurate in reporting. As noted above, each company does this activity differently. The supervisor is responsible for finding and carrying out his or her role in the field reporting system.

10.3.0 Supervisor's Role in Cost Control

The supervisor is the construction representative in the field, where the work takes place. Therefore, the supervisor has a great deal to do with determining job costs.

If the actual costs are at or below the estimated costs, the job is progressing as planned and scheduled, and the company will realize the expected profit. However, if the actual costs exceed the estimated costs, one or more problems may result in the company losing its expected profit and maybe more. No company can remain in business if it continually loses money.

If losses are occurring, the supervisor will be called on to get the costs back in line. How can the supervisor do this? To answer this question, one needs to know some of the major reasons why the costs are running higher than the calculated estimate. Once the reason or reasons are known, the next step is for the supervisor to correct the problem.

Noted below are some of the causes why actual costs exceed estimated costs and suggestions on what the supervisor can do to bring the costs back in line. Before starting any action, however, the supervisor should check with his or her superior to see that the action proposed is acceptable and within the company's policies and procedures.

1. *Cause:* Late delivery of materials, tools, and/or equipment

   Corrective Action: Plan ahead to ensure that job resources will be available when needed

2. *Cause:* Inclement weather

   Corrective Action: Have alternate plans ready

3. *Cause:* Unmotivated workers

   Corrective Action: Implement effective leadership skills

4. *Cause:* Accidents

   Corrective Action: Implement effective safety program

The supervisor should note that there are many other methods to get the job done on time if it gets off schedule. Examples include working overtime or increasing the size of the crew. However, these examples will increase the cost of the job, so they should not be done without the approval of the project manager.

REVIEW QUESTIONS ◆ CHAPTER FOUR, SECTIONS 10.0.0–10.3.0

1. The concept of cost awareness and control means _____.

   a. the supervisor must hire a person in charge of making a monthly budget

   b. the supervisor is responsible for cost reduction, whereas crew members should focus on the actual construction of the project

   c. estimated performance is compared with actual performance

   d. the supervisor must make sure the actual costs of the job are more than the estimated costs

2. It is important that the crew leader be aware of the costs of doing a project for the following reasons *except* _____.

   a. if there are not enough resources on the job site, valuable time can be lost in obtaining them

   b. it can prevent possibly layoffs

   c. the contractor will receive the profit that was estimated before the project began

   d. it allows the contractor to determine if a project is measuring up to its actual cost determined before the project began

3. Crew leaders can help play a role in controlling job costs by _____.

   a. implementing and enforcing safety policies

   b. dismissing unmotivated employees

   c. stopping construction work during bad weather

   d. completing the project at the actual cost that the contractor and client agree upon in the beginning

4. Job costs can be controlled by _____.

   a. giving employees the full responsibility of controlling resources

   b. making sure actual costs at least match estimated costs

   c. using the field reporting system to calculate the labor costs associated with each task on a one-time basis

   d. using the field reporting system to determine the estimated costs

5. If a job is completed one month later than planned, the following job costs are associated with it *except* _____.

   a. overtime payment

   b. money to hire additional workers

   c. money to implement the field reporting system

   d. money to pay for extra crew-hours

SECTION 11

11.0.0 RESOURCE CONTROL

The supervisor's job is to ensure that all jobs are built safely according to the drawings and specifications, within the estimate and schedule. To accomplish this, the supervisor must closely control how resources of materials, equipment, tools, and labor are used.

11.1.0 Control

Control involves measuring performance and correcting deviations from plans to accomplish objectives. Control anticipates deviation from plans and takes measures to prevent its occurrence in the future.

An effective control process can be broken down into the following steps:

Step 1 Establishing standards and dividing them into measurable units.

For example, a standard can be created using experience gained on a typical job, where 2,000 LF of 1¼" copper water pipe was installed in five days. Dividing 2,000 by 5 gives 400. Thus, the average installation rate in this case for 1¼" copper water pipe was 400 LF/day.

Step 2 Measuring performance against a standard.

On another job, the length of the same pipe placed during an average day was only 300 LF. Thus, this average production of 300 LF/day did not meet the average standard of 400 LF/day established above.

Step 3 Correcting operations to assure that the standard is met.

In *Step 2* above, if the plan called for the job to be completed in five days, the supervisor would have to take action to ensure that this happens. If 300 LF/day is the actual average daily rate, it will have to be increased by 100 LF/day to meet the standard.

11.2.0 Materials Control

The supervisor's responsibility in materials control will depend on the policies and procedures of the company. In general, the supervisor is responsible for ensuring on-time delivery, preventing waste, controlling delivery storage, and preventing theft of materials.

11.2.1 Ensuring On-Time Delivery

The supervisor must make sure that the materials for each day's work are on the job site when needed, not waiting until the morning that the materials are needed to obtain them. Rather, the supervisor should use the *one-plus rule*. This means that one week before needing the materials, the supervisor should know where they are, and they must be on the job site one day before they are needed.

The one-plus rule is just a guide. The exact amount of time may need to be adjusted depending upon the situation. To do this effectively, the supervisor must be familiar with the drawings and the activities to be performed. He or she can then determine how many and what types of materials are needed.

If other people are responsible for providing the materials for a job, the supervisor must follow up to make sure that the materials will be available when they are needed. Otherwise, delays will occur as crew members stand around waiting for the materials to be delivered.

11.2.2 Preventing Waste

Waste in construction can add up to loss of critical and costly materials and may result in delaying the job. Because of this, waste has to be prevented. The supervisor must ensure that every crew member knows how to properly use the materials in the most effective manner. The supervisor should monitor the crew to make certain that it is not wasting materials.

An example of materials waste is a carpenter who saws off a piece of lumber from a fresh piece, when the size needed could have been found in the scrap pile. Another example of waste involves installing a fixture or copper pipe incorrectly. The time spent installing the item incorrectly is wasted because the task will need to be redone. In addition, the materials may be lost if the fixture or copper pipe was damaged while being installed incorrectly.

11.2.3 Verifying Material Delivery

In some companies, a supervisor may be responsible for the receipt of materials delivered to the work site. When this happens, the supervisor should require a copy of the shipping ticket and check each item on the shipping ticket against the actual materials to see that the correct amounts were received.

In addition, the supervisor should check the condition of the materials to verify that nothing is defective before signing the invoice. This can be difficult and time consuming because it means that cartons must be opened and their contents must be examined. However, it is necessary because a signed invoice indicates that all of the materials were received unharmed. If the supervisor discovers damaged materials without checking prior to signing the invoice, he or she will be unable to prove that the materials came to the site in that condition.

Once the shipping ticket is checked and signed, the supervisor should give the original or a copy to the superintendent or project manager. The shipping ticket will then be filed for future reference because it serves as the only record the company has to check bills received from the supply house.

11.2.4 Controlling Delivery Storage

Another very important element of materials control is related to where the materials will be stored on the job site. There are two factors in determining the appropriate storage location. The first is convenience. If possible, the materials should be stored near where they will be used. The time and effort saved by not having to carry the materials long distances will greatly reduce the installation costs.

Next, the materials must be stored in a secure area where they will not be damaged. It is important that the storage area suit the materials being stored. For instance, materials that are sensitive to temperature such as chemicals or paints should not be stored in areas that are too hot or too cold. Otherwise, waste will occur.

11.2.5 Preventing Theft and Vandalism

Theft and vandalism of construction materials increase costs because these materials are needed to complete the job; therefore, they must be replaced in order to do so. The time lost because the needed materials are missing adds significantly to the cost. In addition, the insurance that the contractor purchases will increase as the theft and vandalism rate grows.

The best way to avoid theft and vandalism is not to expose anything so that it can be stolen. If all materials delivered to the job site have not been installed by the end of the workday, they should be stored along with any tools in a secure location such as in a construction trailer or truck on the job site. If the job site is fenced or the building can be locked to prevent unauthorized people from entering the premises, the materials can be stored within.

11.3.0 Equipment Control

The supervisor will not be responsible for long-term equipment control. However, the equipment required for a specific job may be the supervisor's responsibility. The first step in determining the kind and amount of equipment needed is to identify the work that will require the machinery or equipment to be transported from the shop or rental yard. Regardless of which worker uses the equipment, it is the supervisor who is responsible for informing the shop where it is being used and seeing that it is returned to the shop upon completion of the task.

Although this may sound like a fairly simple task, it is not. It is common for equipment to lie idle at a job site because the job has not been properly planned and the equipment arrived early. An example is when the wire-pulling equipment and materials arrive at a job site before the conduit has been installed. The result is that this equipment and materials could be used somewhere else instead of awaiting the installation of the conduit. In addition, it is possible that this unused equipment could be damaged, lost, or stolen.

To control equipment usage, the supervisor must ensure that the equipment is operated in accordance with its design and that it is being used within time and cost guidelines. In addition, it is important that equipment be maintained and repaired as indicated by the Preventive Maintenance Schedule. Delaying the maintenance and repairs will result in more costly equipment operation in the long run.

The supervisor is responsible for the proper operation of all other equipment resources, including cars and trucks. Reckless or unsafe operations of vehicles will likely result in damaged equipment and a delayed or unproductive job. This, in turn, could affect the supervisor's job security.

The supervisor should also ensure that all equipment is secured at the close of each day's work in an effort to prevent theft. If the equipment is still being used for the job, the supervisor should ensure that it is locked in a safe place; otherwise, it should be returned to the shop.

11.4.0 Tools Control

Among construction companies, various policies govern who provides tools to employees. Some companies provide all the tools, while others furnish only the larger power tools. The crew leader should find out about and enforce any company policies related to tools.

Tools control is a twofold process. First, the supervisor must control the issue, use, and maintenance of all tools provided by the company. Next, the supervisor must control how the tools are being used to do the job. This refers to tools that are issued by the company as well as tools that belong to the workers.

Using the proper tools correctly saves time and energy. In addition, proper tool usage reduces the chance of damage to the tool being used. Proper usage also reduces injury to the worker using the tool or to nearby workers.

Tools must be adequately maintained and properly stored. Making sure that tools are cleaned, dried, and oiled prevents rust and ensures that the tools are in the proper working order.

In the event that tools are damaged, it is essential that they be repaired or replaced promptly. Otherwise, an accident or injury could result.

Company-issued tools should be taken care of as if they are the property of the user. Workers should not abuse tools simply because they are not their own.

One of the major causes of time lost on a job is the time spent searching for a tool. To prevent this from occurring, a storage location for company-issued tools and equipment should be established. The supervisor should make sure that all company-issued tools and equipment are returned to this designated location after use. Similarly, workers should make sure that their personal toolboxes are organized so that they can readily find the appropriate tools. In addition, they should return their tools to their toolboxes when they are finished using them.

Studies have shown that the key to an effective control system for tools is limiting the following:

- The people allowed access to stored tools
- The people held responsible for tools
- The ways in which a tool can be returned to storage

11.5.0 Labor Control

The hardest aspect of any job is to control labor. Improving labor productivity results in greater savings and lower construction costs. Many studies have shown that construction personnel are only 35 to 40 percent efficient; that is, in one hour of a typical day, only 35 to 40 percent of the time is spent doing the assigned task. The remaining 60 to 65 percent of the time is devoted to financially unproductive or indirect tasks.

Time and motivation studies indicate that a typical worker's day consists of the following:

- Receiving instructions
- Reaching for materials, tools, and equipment
- Searching for materials, tools, and equipment
- Walking with materials, tools, and equipment
- Doing the work of his or her trade
- Working on assignments other than those within his or her trade
- Waiting for other workers to complete their tasks
- Waiting for materials, tools, and equipment
- Walking and loading
- Standing idle for no apparent reason
- Talking about matters other than work
- Delaying for personal reasons
- Planning, discussing, and studying the layout

To improve productivity, the supervisor needs to reduce the time that the worker is not being productive.

Studies also reveal that when a worker does not do a reasonable day's work, it is because there is a lack of one or more of the following:

- Information
- Work (or too many interferences)
- Tools
- Motivation
- Ability
- Desire

An effective program consisting of planning, scheduling, and supervision can correct the first three issues. An effective leadership and training program can correct the last three issues.

The kind and amount of labor will vary from job to job. The supervisor may not be expected to plan the labor needed for the entire job. However, he or she may be asked to offer recommendations based on the quantity and character of the work to be completed in the immediate future.

Frequently, the amount of work to be completed is stated in percentages. For example, 30 percent of the job has to be completed within two weeks. Once the percentage is established, it is necessary to determine the amount of unskilled and skilled labor required to meet the objective.

To estimate the amount of labor required, the supervisor will have to determine the following two things:

- The kind of work
- The estimated amount of time needed to complete the tasks

The supervisor should be able to identify the kind of work immediately. However, he or she may have to rely on past experiences to determine the amount of time needed to complete the task(s).

There are a few essential labor-related practices that should be followed in an effort to complete the job efficiently. They are as follows:

- Ensure that all workers have the required resources when needed
- Ensure that all personnel know where to go and what to do after each task is completed
- Make reassignments as needed
- Ensure that all workers have completed their work properly

PARTICIPANT EXERCISE H

1. How does your company minimize waste?

   _____

   _____

   _____

   _____

   _____

2. How does your company control the use of small tools on the job?

   _____

   _____

   _____

   _____

   _____

   _____

3. List five ways that you feel that your company could control labor.

   _____

   _____

   _____

   _____

   _____

   _____

   _____

   _____

   _____

   _____

PARTICIPANT EXERCISE H ◆ PROJECT CONTROL

SECTION 12

12.0.0 PRODUCTION AND PRODUCTIVITY

Pay for time on the job is based on the work that needs to be performed. Using time productively results in more work being completed. Many savings that we have discussed are directly related to the effective usage of time.

Planning ahead is basic to a high rate of productivity. By thinking ahead, a supervisor can eliminate wasted time. The supervisor and crew will be prepared to perform the required work, and they will have the materials, tools, and equipment required to do so available.

The time on the job should be for business, not for taking care of personal problems. Anything not work-related should be handled after hours, away from the job site. Planning after-work activities, arranging social functions, or running personal errands should be handled after work or during a break.

Organizing fieldwork can save time. The key to effectively using time is to "work smarter", not necessarily harder. For example, most construction projects require that the contractor submit a set of as-built drawings at the completion of the work. These drawings describe how the materials were actually installed. The best way to prepare these drawings is to mark a set of working drawings as the work is in progress. That way, pertinent details will not be forgotten and time will not be wasted trying to remember how the work was done.

Supervisors are very familiar with the terms production and productivity. Production is the amount of construction put in place. It is the quantity of materials installed on a job, such as 1,000 linear feet of waste pipe installed in a given day. On the other hand, productivity depends on the level of efficiency of the work. It is the amount of work done per hour or day, by one worker or a crew.

Production levels are set during the estimating stage. The estimator determines the total amount of materials to be put in place from the drawings. After the job is complete, the actual amount of materials installed can be assessed, and the actual production can be compared to the estimated production.

The quantity of materials used should be equal to or less than estimated. If it is not, either the estimator has made a mistake, undocumented changes have occurred, or rework has caused the need for additional materials. Whatever the case, the supervisor should use effective control techniques to ensure the efficient use of materials.

Productivity relates to the amount of materials put in place by the crew over a certain time period. The estimator uses company records during the estimating stage to determine how much time and labor it will take to place a certain quantity of materials. From this information, the estimator calculates the productivity necessary to complete the job on time.

For example, it might take a crew of two people 10 days to paint 5,000 square feet. The crew's productivity per day is obtained by dividing 5,000 square feet by 10 days. The result is 500 square feet per day. The supervisor can compare the daily production of any crew of two painters doing similar work with this average, as discussed in *Planning Resources* in this chapter.

In addition, information gathered from the field through field reporting will allow the home office to calculate the actual productivity and compare it to the estimated figure. If the actual production was below that estimated, actions would have to be taken during construction to improve the productivity. In this way, the job can be built at or below the estimated cost and within the scheduled time.

REVIEW QUESTIONS ◆ CHAPTER FOUR, SECTIONS 1.0.0–12.0.0

1. One of the three phases of a construction project is _____.

   a. the development phase, in which design professionals produce drawings and other plans which are then shown to the contractor and to other workers

   b. the construction phase, in which architectural and governmental agencies give a final inspection of the design, adherence to codes, and materials used

   c. the planning phase, in which estimates of all expenses are made

   d. the construction phase, in which the owner selects a contractor

2. One of the three types of project delivery systems is _____.

   a. design-build, in which the owner designs how something is to be constructed

   b. general contracting, in which the contractor is in charge of the design

   c. construction management, in which the contractor builds according to the owner's design

   d. general contracting, in which the owner hires a contractor

3. Planning can be defined as all of the following *except* _____.

   a. deciding how to complete a project

   b. breaking down the project into tasks

   c. developing a step-by-step procedure

   d. visiting the job site a few hours before the project is to begin, in order to check the availability of equipment

4. One reason it is important to plan is _____.

   a. it will help in completing the job at a lower cost

   b. it will minimize the unexpected

   c. it will prevent all delays from occurring

   d. it will guarantee a successful project

5. The planning process includes _____.

   a. setting objectives that identify the work to be done, even though workers might not approve of them

   b. breaking down large jobs into subtasks

   c. visiting the job site to see how well the project is going

   d. involving the supervisors, who typically have the responsibility of establishing goals for their workers

6. The supervisor should plan what tools will be used for a project by _____.

   a. letting workers decide which tools will be used for the job

   b. letting workers know who will provide the tools

   c. asking the workers if they know the safe operation of the tools that they will be using

   d. letting the workers set the controls for the tools required

7. A job diary should typically indicate _____.

   a. items such as job interruptions and visits

   b. only the major incidents of the day

   c. the estimated time for each job task related to a particular project

   d. the supervisor's ideas for improving employee morale

8. One example of a direct cost when building a job is _____.

   a. office rent

   b. job-site telephones

   c. equipment

   d. utilities

9. All the following are ways for a contractor or construction company to earn a profit *except* _____.

   a. the supervisor should follow job specifications as closely as possible

   b. the contractor's main office should perform a quantity takeoff

   c. the supervisor should not allow for much deviation from the schedule during the course of the project

   d. the supervisor should attempt to reduce direct costs

10. The project estimate is prepared _____.

   a. just after the project begins

   b. after the project is completed

   c. just before the project is going to end

   d. before a project begins

11. Among the documents needed for preparing a project estimate is _____.

   a. any drawing or specification in order to complete a Summary or Pricing Sheet

   b. a Recapitulation Sheet, which basically sums up the project

   c. a Quantity Takeoff Sheet, which indicates how many materials are needed for a job

   d. the Estimated Profit Sheet, which serves as the final document in a project estimate

12. An example of overhead is _____.

   a. utilities

   b. labor

   c. material

   d. tools

13. The process of determining how much a job will cost is known as _____.

   a. the estimate

   b. the material quantity

   c. the direct cost

   d. the indirect cost

14. Scheduling involves all of the following *except* _____.

   a. making use of a bar chart

   b. determining the estimated cost

   c. determining on what date specific tasks will end

   d. determining the amount of materials required to complete the job

15. One of the tasks required to develop a schedule is to _____.

   a. determine during the first step of the scheduling process how long each activity will take

   b. make an initial list of all the activities required to complete the job, noting who will perform each individual task

   c. allow time for inspections

   d. choose a start time for each job activity but, to allow for flexibility, avoid making guesses about how long each activity might last

REVIEW QUESTIONS ◆ CHAPTER FOUR, SECTIONS 1.0.0–12.0.0
(Continued)

16. The difference between estimated costs and actual costs is that _____.

   a. estimated costs are determined just after a project gets underway

   b. actual costs are the costs, agreed upon in a contract, necessary to complete a job

   c. estimated costs are adjusted at the end of a project if they do not match the actual costs

   d. estimated costs serve as a type of cost guide to the contractor throughout the project

17. One of the steps of the control process typically involves _____.

   a. establishing a network schedule

   b. determining if performance was up to established standards

   c. getting an estimated cost for a project

   d. figuring out the actual cost of a project

18. The supervisor should control the use of resources such as _____.

   a. tools, which does not include those tools that actually belong to the workers

   b. labor, which can typically involve providing training

   c. equipment, which involves equipment control for multiple jobs over a long period of time

   d. materials, by having crew members sign a statement saying they will not waste materials

19. The supervisor generally controls the use of materials by doing the following *except* _____.

   a. assuring on-time delivery, in order to avoid delays

   b. controlling delivery storage, keeping materials near where they will be used

   c. preventing theft and vandalism using the *one-plus rule*

   d. verifying material delivery and checking for deficient items

20. To improve both production and productivity, it is essential to _____.

   a. eliminate rest breaks

   b. after a project is fully completed, report any essential details regarding how the materials were used for the project

   c. require overtime

   d. control the use of materials so that they are used efficiently

ADDITIONAL RESOURCES ON THE WEB

American Medical Association (AMA)	www.ama-assn.org
American Society for Training and Development (ASTD)	www.astd.org
Architecture, Engineering, and Construction Industry (AEC)	www.aecinfo.com
CIT Group	www.citgroup.com
Construction Work Development Center	www.construction-work.org
Equal Employment Opportunity Commission (EEOC)	www.eeoc.gov
Jamestown Area Labor Management Committee (JALMC)	www.jalmc.com
Knowledge Center's Manager's Toolkit	www.knowledgecenters.versaware.com
National Association of Women in Construction (NAWIC)	www.nawic.org
National Census of Fatal Occupational Injuries (NCFOI)	http://stats.bls.gov/news.release/cfoi.toc.htm
National Center for Construction Education and Research	www.nccer.org
National Institute of Occupational Safety and Health (NIOSH)	www.cdc.gov/niosh
National Safety Council	www.nsc.org
Occupational Safety and Health Administration (OSHA)	www.osha.gov
Society for Human Resources Management (SHRM)	www.shrm.org
United States Bureau of Mine Safety and Health Administration (MSHA)	www.msha.gov
United States Census Bureau	www.census.gov
United States Department of Labor	www.dol.gov
USA Today	www.usatoday.com

The NCCER makes every effort to keep these textbooks up-to-date and free of technical errors. We appreciate your help in this process. If you have an idea for improving this textbook, or if you find an error, a typographical mistake, or an inaccuracy in NCCER's Contren™ textbooks, please write us, using this form or a photocopy. Be sure to include the exact module number, page number, a detailed description, and the correction, if applicable. Your input will be brought to the attention of the Technical Review Committee. Thank you for your assistance.

Instructors – If you found that additional materials were necessary in order to teach this module effectively, please let us know so that we may include them in the Equipment/Materials list in the Instructor's Guide.

Write: Curriculum Revision and Development Department
National Center for Construction Education and Research
P.O. Box 141104, Gainesville, FL 32614-1104

Fax: 352-334-0932

E-mail: curriculum@nccer.org

Craft _____ Module Name _____

Copyright Date _____ Module Number _____ Page Number(s) _____

Description _____

_____

_____

_____

(Optional) Correction _____

_____

_____

(Optional) Your Name and Address _____

_____

_____

Index

Index